Warren Brown is an award-winning cartoonist and writer for the *Daily Telegraph* and *Sunday Telegraph* newspapers and is well known as a television and radio presenter.

He has a love of the outback, adventure and Australian history. To commemorate the construction of the Stuart Highway in the Northern Territory and to mark the fiftieth anniversary of the end of World War II, he co-conceived *Back to the Track* – a pilgrimage for veterans, which saw 100 World War II vehicles freighted to Alice Springs by rail and driven in convoy to Darwin. He co-conceived the re-creation of the Peking to Paris race of 1907 using five original cars and with co-driver Lang Kidby was awarded the *Australian Geographic* Spirit of Adventure Award.

Since 2006 Warren has been the overnight host at Gallipoli on Anzac Day and in 2010 was appointed a member of the National Commission on the Commemoration of the Anzac Centenary. He lives in rural New South Wales with his wife Tanya and son Oliver.

LASSETER'S GOLD

WARREN BROWN

hachette
AUSTRALIA

For the men of the Central Australian Gold Exploration Expedition, 1930–31.

hachette
AUSTRALIA

Published in Australia and New Zealand in 2015
by Hachette Australia
(an imprint of Hachette Australia Pty Limited)
Level 17, 207 Kent Street, Sydney NSW 2000
www.hachette.com.au

Text, line illustrations and maps copyright © Warren Brown 2015

This book is copyright. Apart from any fair dealing for the purposes of private study, research, criticism or review permitted under the *Copyright Act 1968*, no part may be stored or reproduced by any process without prior written permission. Enquiries should be made to the publisher.

National Library of Australia
Cataloguing-in-Publication data:

Brown, Warren, author.
Lasseter's gold/Warren Brown.

978 0 7336 3160 3 (paperback)

Lasseter, Harold Bell, –1931.
Lasseter's Reef (Legendary place).
Prospecting – Australia, Central – History.

919.42

Cover design by Saso
Cover photograph courtesy of Michael Koch
Text design by Bookhouse, Sydney
Typeset in Simoncini Garamond
Printed and bound in Australia by Griffin Press, Adelaide, an Accredited ISO AS/NZS 14001:2009 Environmental Management System printer

MIX
Paper from responsible sources
FSC® C009448

The paper this book is printed on is certified against the Forest Stewardship Council® Standards. Griffin Press holds FSC chain of custody certification SGS-COC-005088. FSC promotes environmentally responsible, socially beneficial and economically viable management of the world's forests.

CONTENTS

	Preface	vii
1	The Treachery Act	1
2	Hoodoo	6
3	All bets are off	10
4	'An out of the ordinary suggestion ...'	13
5	The end justifies the means ...	25
6	Gold	30
7	The Bicycle Bushman	39
8	Machines	47
9	'A fellow named Burlington ...'	54
10	'Desirous of going to Central Australia ...'	58
11	Planes, trains and automobiles	66
12	'We draw the colour line here ...'	74
13	Strange behaviour	77
14	Preparations	84
15	Archie Giles	96
16	The Dashwood	106
17	The airstrip	114
18	Haasts Bluff	120
19	Death trap	132
20	Crash through or crash	136

21	Ilpbilla	148
22	'Approaching my country'	157
23	Who was Lasseter?	165
24	New directions	176
25	News	196
26	*Golden Quest II*	209
27	Looking for the dogger	220
28	Rocks and hard places	234
29	The Sydney crowd	251
30	The stranger in town	254
31	The search for the searchers	267
32	Bob Buck	272
33	Strange turns	280
34	Torn apart	296
35	Awkward questions	302
36	Tailings ...	316

Afterword	325
Acknowledgements	329
Sources	335

PREFACE

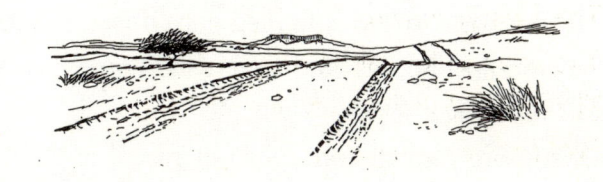

There is a remarkable, intriguing photograph taken on the track north of Alice Springs during the winter of 1930 – a group portrait of six men posed uncomfortably before an all-pervasive camera. Their body language suggests this cluster of disparate individuals might have just been apprehended for some sort of crime – indeed, the group appears as though they could be a band of early twentieth-century wild-west outlaws, somewhat like Butch Cassidy's Hole-in-the-Wall-Gang. There is nothing collegiate about the men in the photograph – it seems the only thing they appear to share is a mutual discomfort in being photographed together. Only one man is looking directly at the lens – his expression is one of resigned determination. The others gaze in different directions – to the horizon, or at their feet – anywhere but towards each other. The photograph suggests these men have probably been assembled here begrudgingly, and that as soon as the picture is taken, they'll no doubt wander away in different directions.

But this photo is not of a police line-up – it is in fact a visual record of the members of the Central Australian Gold Exploration Company expedition. A mission devised to search for and locate a vast deposit of gold said to be hidden somewhere in the middle of Australia, that stretched for mile after treasure-filled mile. It would be the biggest gold find in history – and the man sitting at the forefront of the photograph wearing a homburg hat knew precisely where it was. His name was Harold Bell Lasseter, a fifty-year-old prospector who had stumbled onto this gold 'reef' thirty years before and now he was guiding a cashed-up expedition into the wilds to rediscover it.

Yet this sombre photograph does not reveal any glimmer of the enthusiasm one would expect from a group of men who would become instant millionaires on the reef's discovery. There is more to the story. When the picture was taken they were only just beginning their search for the treasure, but already the cancer of doubt and mistrust had set in.

Between July 1930 until February 1931 the search for Lasseter's gold reef traversed an area west of the Alice, roughly the size of the state of Victoria. It should have been a rattling tale of high adventure resulting in the eventual discovery of his El Dorado – but somehow, things went wrong. Badly wrong.

What emerged from the desert west of Alice Springs in the early 1930s was a mystery that confounds people to this day. For everyday Australians who anxiously followed the unfolding story of the expedition's troubles through newspaper articles, the tale of Harold Lasseter and his gold reef changed from a tale of intrigue to become a mystery, before evolving into a myth of enormous proportions – forever cemented into Australian folklore.

This book tells that *tale* of Harold Bell Lasseter and the events that befell the Central Australian Gold Exploration expedition of 1930–31. With Lasseter employed as a guide, a team of men went in search of a vast reef of gold the prospector said he'd found somewhere in the centre of the Australian continent many years before. Precisely what transpired out near the border of the Northern Territory and Western Australian almost a century ago will never be known. The expedition leader Fred Blakeley did not maintain any kind of official diary or log. All the participants are now long dead and their surviving testaments about what happened are all written with an agenda.

There are countless paths that could be followed when telling the Lasseter story. It's tricky territory in which to tread – for the best part of a century the lure of finding gold has seen Lasseter's every move and every written word forensically analysed in the hope of finding any clue as to the reef's whereabouts. To exhaust every theory, every rumour, every lead, every dead-end and every detail regarding the eccentric Harold Lasseter and the hunt for his treasure would require the publication of a book many times this size. I have tried to capture the essence of what transpired and conclude with some observations. While I have not invented any part of the story, it is important to bear in mind that some of the accounts from which I have shaped the narrative could well be unreliable – written down almost a century ago to perhaps obscure what really may have happened. In this regard I viewed the Lasseter story as a tale – where truth has sometimes been interwoven with . . . well . . . with whatever someone wished to place on record. Which makes it all the more intriguing – it is a tale with some very surprising twists.

I have tried to place the story within the framework of its time – it is important to understand the desperate search for Lasseter's gold was very much fuelled by the calamity of the Great Depression and the scramble to revive personal fortunes, as well as those of the nation.

The narrative for this book has been derived from many sources. Some were from first-hand accounts written soon afterward and years later, some from documents located within the files of the Mitchell Library of NSW, the National Archive, the National Library and various other cultural institutions. Some accounts are taken from the writing of historians – professional and amateur – and many from digitised newspapers and magazines, thanks to the never-ceases-to-amaze National Library of Australia's Trove archive. All dialogue within the narrative is taken from written accounts found in books, newspaper articles, documents and film – none is invented by me. I've tried to let the characters speak for themselves – they tell the story and there is a joy in hearing their long-lost voices using the vernacular of almost a century ago.

There are some decisions I've made regarding the spelling of particular names. For example the airstrip at Ilpbilla is sometimes written as Ilbilla, Ilbilba and Ilipi, among other variations. To my understanding, all are acceptable spellings – I've decided on Ilpbilla as this is the name used in the report to government by explorer Donald McKay when the site was cleared for aircraft only months before Lasseter's expedition arrived.

Throughout the story there are noticeable, jarring inconsistencies with given times, dates and measurements – the length of Lasseter's gold reef for example, varies from being anywhere from seven to fifteen miles long. The year he discovered it is also open to speculation, as are the distances he'd said he'd travelled. These inconsistencies are all part of the Lasseter tale – generally most people turned a blind eye to them, choosing not to question anything that would raise doubts to the veracity of the story, but eventually they become a problem.

Further, there is a charm in just how glacially paced the world was back in 1930. In the twenty-first century it is almost inconceivable for someone to suggest, 'Well, you wait here for three weeks while I travel by camel two hundred miles . . .' Time and distance seemed to mean very little – except for the race to beat the oncoming summer.

Lasseter was once a name that could instantly spark fierce arguments, a name that rallied countless gold-hungry expeditions to plunge into Australia's dead heart, a name which for decades engendered strong opinions and robust convictions. But in time, the storm of hope and despair and anger and loss and the frustration Harry Lasseter's name evoked has almost passed – modern Australians who've heard of him recall what they know of the Lasseter story as something of a quaint and perhaps a misguided aberration in Australia's history.

I learned for some the *notion* of Harold Lasseter lay somewhere filed away in the Australian subconscious inbox, along with Burke and Wills and Ludwig Leichhardt and all those other pioneering types who, for whatever reason, wandered out into the middle of Australia only to come to a sticky end. When researching the book, friends I canvassed had *heard* of Lasseter but, interestingly, when I asked them what they *knew* about Lasseter, were surprised – perhaps even shocked – to learn Lasseter's journey took place in 1930, involving the use of trucks and cars and aircraft and radios. Quite a few people were under the impression the Lasseter story resided somewhere in the latter half of the nineteenth century – a world and a time too remote for modern Australians to readily connect. But the realisation Lasseter's search for gold took place between the First and Second World Wars, utilising the very latest in twentieth-century technology seemed to spark genuine surprise and intrigue.

Indeed, there is a modern, perhaps timeless relevance to the story of Harold Lasseter and the pursuit of his lost gold reef – the indescribably priceless treasure always just out of arm's reach.

What began as the sort of yarn one might hear around a campfire or in a pub, overnight became a full-blown publicly subscribed expedition, which ultimately grew into a grim parable worthy of the Old Testament.

But what is truly remarkable about the Lasseter story is that his elusive gold reef runs a distant second place to the fascinating people who are drawn to him. The real treasure in Lasseter's tale is the coterie of self-serving and devious individuals who are there only to take advantage of the strange and inscrutable gold prospector. Their immorality is at times staggeringly difficult to believe – the toxicity of gold fever exposing dark and sinister traits in people.

The story of the elusive gold reef would inspire generations of gold-fevered prospectors to head out on the hunt for where Lasseter left off – no doubt proving the legend of his astonishing find to be true. But the reality of what transpired back in 1930 was something quite different. There was little of the rollicking adventure of H. Rider Haggard's *King Solomon's Mines* here. What took place was even more interesting.

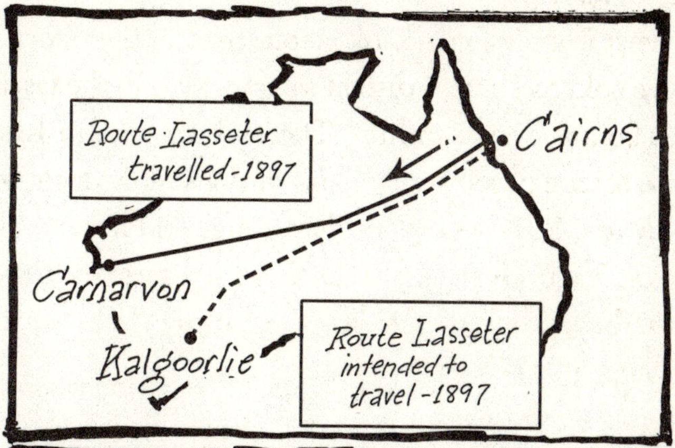
Route Lasseter
travelled-1897
Cairns
Carnarvon
Kalgoorlie
Route Lasseter
intended to
travel-1897

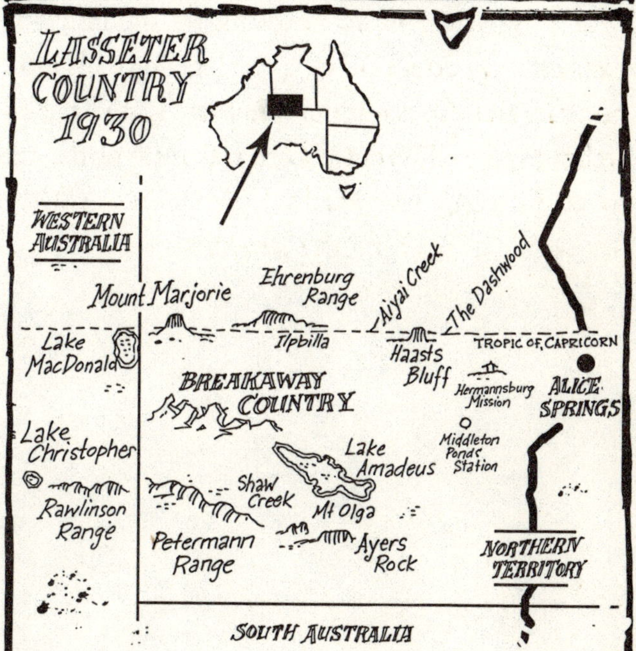
LASSETER
COUNTRY
1930
WESTERN
AUSTRALIA
Mount Marjorie
Ehrenburg
Range
Aiyai Creek
The Dashwood
Lake
MacDonald
Ilpbilla
Haasts
Bluff
TROPIC OF CAPRICORN
Hermannsburg
Mission
ALICE
SPRINGS
BREAKAWAY
COUNTRY
Lake
Christopher
Lake
Amadeus
Middleton
Ponds
Station
Shaw
Creek
Mt Olga
Rawlinson
Range
Petermann
Range
Ayers
Rock
NORTHERN
TERRITORY
SOUTH AUSTRALIA

'Those whom the gods wish to destroy they first make mad . . .'
– attributed to Euripides

I

THE TREACHERY ACT

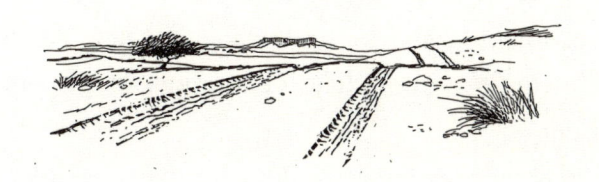

WESTMINSTER, LONDON, 1940

The beginning of May 1940 was the prelude to what Winston Churchill would gravely describe as Britain's 'darkest hour'. At month's end, more than 330,000 shattered and demoralised British and French soldiers would be evacuated from the beaches at Dunkirk. The might of a rolling German army would amass at the edge of the English Channel as if they were a pack of restrained attack dogs, at any moment unleashed to make the leap and go in for the final kill. Britain was bracing itself for the inevitability of a German occupation.

There had been much talk of German spies already operating in Britain, and the counter-intelligence service MI5 was dealing with an increasing number of arrests. The race was on to locate and decommission these espionage agents as swiftly as possible. There had been some success in rounding up bona fide Nazi spies. The outcome, if found guilty under the new Winston

Churchill-instituted Treachery Act, was execution by hanging or firing squad.

Early in May, information had come to light suggesting there was something unusual unfolding within an Air Raid Precautions (ARP) crew based at Dolphin Square in Westminster. The suggestion was that the Civil Defence Rescue and Shore Party had been infiltrated by fascist operatives – in particular, 'blackshirts' loyal to Britain's chief pro-Nazi, Sir Oswald Mosley, who within a matter of weeks would be interned under new defence regulations allowing the immediate detainment of suspected Nazi sympathisers.

Dolphin Square, an enormous private apartment block on the edge of the Thames described at its completion in 1937 as 'the largest self-contained block of flats in Europe', was home to an inordinate number of lords and members of parliament, making it a prime target for anyone wishing to eliminate a significant portion of the British political machine in one swoop. And it is most likely that the information revealing the ARP crew's operations came from a branch of MI5 known as B5(b) – a section solely responsible for infiltrating its own agents into potentially subversive groups, which happened to have its office within the Dolphin Square complex.

MI5 learned that this small subversive group operating within Civil Defence was headed by a German-born agitator who had been less than discreet in his plotting, boasting to undercover agents of his proposed plans to engage in some sort of sabotage.

British Intelligence rounded up the ARP crew for cross-examination, only to discover they had picked up someone who seemed to be a Nazi insurgent. Thirty-four-year-old Albert Paul Johns might well have been a Nazi spy fresh from Central Casting – he was handsome, cold, arrogant, confident, and totally dismissive of his captors. He spoke perfect English with a clipped German accent and cited his Nazi past openly, confessing to his actions.

Johns had unwittingly bragged to covert agents of his involvement in the November Pogrom of 1938 – otherwise known as the Kristallnacht – and of his training within Nazi instruction schools and very likely the SS. A report signed off by the director general of MI5, Major-General Vernon Waldegrave Kell, revealed that Johns had arrived in London only a few weeks before the outbreak of war and had sought to join Mosley's fascists as they 'gave him the best opportunities for finding conscious or unconscious – but anyway active – support for his activities in the interest of Hitler'.

Yet there was something of an unsettling air of bravado about this strange young German – for all his self-proclaimed Nazi pedigree it didn't seem that he was much of an espionage mastermind. Instead his antics were confined to little more than the waste of materials and the destruction of property. But as it transpired, Johns's history had other aspects that made him all the more intriguing.

Johns possessed a British passport, but his interrogators were surprised to learn that at some point he had renounced his German citizenship to become a naturalised Australian. Just how a Nazi saboteur in London had acquired Australian citizenship was no doubt going to make for an interesting story. As it transpired, he had spent several years wandering through the wilderness of Central Australia – as a dog shooter, a cameleer and a prospector – and had later performed odd jobs in Brisbane and Sydney. Much to the surprise of his interrogators, Johns announced that he was intending to return to Australia – to Alice Springs in fact – yet the way his statement was unfolding he was heading for the gallows.

On 20 May, a letter marked secret was sent to Lieutenant Colonel H. E. Jones of the No. 1 Secretariat in Canberra announcing that British Intelligence had picked up Johns – an Australian national – and that he was intending to return to Australia. The letter outlined Johns's espionage activities, his time in Nazi Germany and his plans to return to Alice Springs.

The reply from Australia was unequivocal, they didn't want him back. It reported that several years earlier, Johns had spent six months in Brisbane's Boggo Road Gaol for car theft and upon release was deported to Germany as an undesirable. As far as Australia was concerned, it would revoke his naturalisation certificate and was more than happy for him to stay behind bars in England. This probably saved Johns from execution. By war's end, some 16 in his predicament would be shot or hanged, but he would spend the rest of the war in a British gaol.

Yet the one thing that MI5 interrogators didn't know about their Nazi 'spy' was that he possessed another interesting story altogether. He was a vital piece in a puzzle that confounds the Australian nation to this day. Ten years earlier, in the dead heart of the Australian continent, Johns had set out to ingratiate himself with what was at that time the most audacious gold exploration expedition Australia had ever seen. With five camels and some Aboriginal guides, the young German had wandered through merciless desert hundreds of miles west of Alice Springs to try to head off a spectacularly well-publicised party of prospectors in search of what was promising to be the greatest gold find in history. They were looking for a fabled gold 'reef' 15 miles long where nuggets lay on the ground 'as thick as plums in a pudding' – or so popular theorists claimed. A prospector, Harold Bell Lasseter, insisted that he had discovered the reef decades before, and he had convinced enough influential people to mount an expedition in search of his prize. However, within a few months the desert foray had disbanded acrimoniously, the party members returning to Sydney demoralised, angry and empty-handed.

Except for Lasseter. He had stayed on, stumbling across the desert in search of his prize. With him was a strange young German dingo scalper who would ultimately return to Alice Springs alone. Albert Paul Johns – the Nazi spy and saboteur, the ex-convict, the man deemed by the Australian government an 'undesirable', and who was considered as a candidate for execution in Britain – was the last white man to ever see the prospector, Harold Bell Lasseter, alive.

2

HOODOO

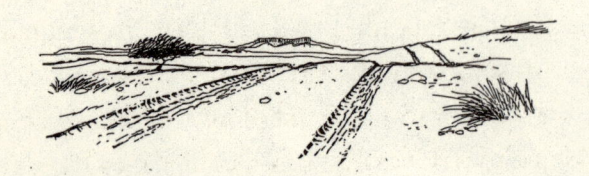

TEN YEARS EARLIER – ALICE SPRINGS, 1930

The hot, dark public bar at the Stuart Arms Hotel in Alice Springs was hardly an inviting place for travellers and 'new-chums'. Conversation between the local drinkers and itinerant travellers within the dusty confines of this lightless, random-stone building was rare to non-existent. 'Blow-ins', as strangers were disparagingly called, were rarely spoken to by the cattlemen and cameleers, who kept to themselves, sipping quietly on flat brown ale. There were gold prospectors too, who had made their way south from Tennant Creek, following the overland telegraph line – hard men, often unpleasant, whose lives creaked along in the wilds of the Never Never with little regard for the law.

But sometimes you might overhear the reticent, begrudging discourse about a phenomenon peculiar to this part of the world. It was about what lay west of the Alice. A story that there was something of a curse – or a 'hoodoo' as the locals described it – plaguing the Petermann Ranges located out in the remote south-western pocket of the Northern Territory.

The Petermanns are the gnarled remains of an ancient series of interconnecting mountains stretching for several hundred miles – remnants of worn-away granite and sandstone peaks that once soared to heights rivalling the Himalayas. By some, the Petermanns were considered a geological mystery: the violent continental collision between India and western Australia some 550 million years ago should by all rights see the mountain range running north–south, but the Petermanns strangely lie east–west.

To the knowing, it was a vicinity that was out of the question. Since explorer Ernest Giles first laid eyes on the mountains in 1872, the handful of white Australians and Afghan cameleers who had ventured into the wastelands west of Alice Springs would ultimately return to the outpost town writing off the Petermanns as impenetrable – extraordinarily remote, labyrinthine, bereft of water and inhabited by mysterious and aggressive Aboriginal tribes, some of whom had never experienced contact with Europeans before.

Yet those who had braved the journey out towards the Western Australian border brought back stories of vast fields of quartz, and talk grew of the possibility of finding gold. Perhaps, way out in the worthless backblocks of the Never Never, there was an El Dorado waiting to be unearthed. Certainly, there was something about this mysterious mountain range that captured the imaginations of those infected with the incurable malaise – gold fever.

The allure of finding gold tempted many, particularly in the desperate years of the Great Depression, but simply travelling to the Petermanns required serious commitment and determination. As the old-timers would tell you, in the summer months the question of finding water would be a key deterrent. There was none. Every drop of water would need to be carried in and accounted for. For those who were still willing to try, it would be a 430-mile camel journey south-west from Alice Springs travelling through sharp spinifex-covered desert, incorporating the daunting challenge of crossing Lake Amadeus, a pitiless 390-square-mile salt lake. And once at the foothills, there was the risk of encountering malevolent, spear-throwing locals. The likelihood of finding death was more on the cards than finding a mother lode.

And if that wasn't enough to deter fortune hunters, those who were still contemplating the journey didn't need to look far to discover that the story of the 'hoodoo' seemed to have some provenance.

In 1906, 33-year-old Frank George, a government prospector, led a gold-finding expedition to the foothills of the Petermanns, where the party was attacked by a tribe of Aborigines, one of the expedition members being fatally speared where he slept, and another speared through the eye. As George attempted a return journey to Alice Springs with his wounded companion, the searing outback heat saw the demise of three of his camels, one through lack of water and two through disease. Only five miles from the town, the expedition leader himself succumbed to dysentery. As far as anyone who had ever heard of the mountains was concerned, the area was cursed. It would be another 20 years before anyone attempted any further exploration back out there, the *Adelaide Mail* in 1926 reminding its readers that since that ill-fated party had visited in 1906, the 'veil of silence had been drawn closely over the Petermann Range area . . .'

3
ALL BETS ARE OFF

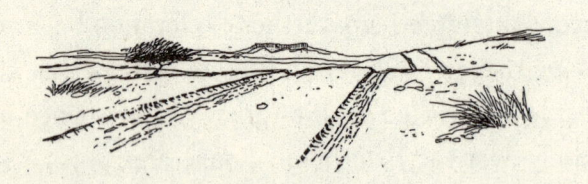

Time and the inevitable wisdom of hindsight can delude us into thinking we understand people's motivations and reactions at a particular moment long ago. But it is difficult to truly comprehend how or why decisions were made in a catastrophic instant; and afterwards, when all the information comes to light and is pieced together, we can sometimes adopt a judgemental – even pious – attitude. To try to comprehend a person's thinking a century ago is particularly difficult. It requires embracing the circumstances of the moment, putting aside an outcome we know in hindsight.

When Australian soldiers sailed for European battlefields in late 1914, they were nervous that the fantastic adventure they were embarking on might be finished before they got there. They were not to know that the war would drag on for four years, and that when the Western Front degenerated into a bloody impasse – a 'war of attrition' – they would be faced with the possibility that they were fighting a war that might never finish. The same could be said for the onset of the Great Depression in 1929: those living through it might conceivably believe it would never end.

Eleven years after the signing of the armistice document declaring the end of World War I, the surviving Australian volunteers who had returned from the catastrophes of Gallipoli and the Western Front – attempting to press-fit their lives into a form of normality they were no longer familiar with – were suddenly confronted with a different kind of war, a global economic catastrophe the likes of which no-one had ever seen.

The arrival of the Great Depression in 1929 threw to the wind everything the returned serviceman understood to be fair and just. In an instant, jobs, savings, security had gone. For many Australians, blind faith in what were once rock-solid institutions – government, political systems, banks, the church – was smashed to the ground along with Wall Street's collapse. Newspapers told of bankrupted New York financiers throwing themselves off skyscrapers; a photograph depicted an expensive Packard car, its despondent owner standing alongside a handwritten sign: 'Must Sell – $100'.

The clear failure of the ferociously boiled-dry and gasket-blown capitalist phenomenon of the roaring twenties had wrecked millions of lives across the Western world, and people began looking for answers. Now they were even questioning the worth of democracy. Indeed, what good was a democratically elected government if it was unable to guarantee food on the table for its constituents? All bets were off.

For many, the time was ripe for considering radical alternatives to bring Australia back on track. Despite all the scaremongering about communism, one only had to look at Russia for a success story. Certainly, they had gone through that unpleasant patch in murdering the Tsar and his family, speakers in Sydney's Domain would argue, but today, everyone in Communist Russia had a job and pay in their pocket. It was hard to argue. Newspapers in the United States, where 13 million people were out of work, carried advertisements from Russia's Soviet government calling out for US skilled labour – an open invitation for Americans to emigrate with their families to the Soviet Union, where the worker would no longer toil in 'dark, gloomy shops with dim, yellow lamps – there will be clean halls with great windows and beautiful tile doors . . . where you'll work seven hours per day instead of eleven, where new factories will be surrounded by flowers with cafeterias and libraries for the workers, nurseries for the children – even swimming pools'.

The onset of the Depression provided an open microphone for other ideas as well. The Light Horse veteran and Rhodes scholar Wilfrid Kent Hughes, Victorian president of the Young Nationalists and member of State Parliament, in 1929 decried parliamentary democracy as 'a curse', instead embracing the principles of Mussolini's fascism – praising the Italian dictator's efforts as 'innovative', with the suggestion his ideals could be 'adapted to Australian parliamentarianism'. Nazism was only around the corner.

The Depression had unleashed an ideological free-for-all, but ideology didn't put money in your pocket. It was all very well to field worthy debate about what political path Australia should decide upon to find its way out of the global economic mire, but what about the individual? After all, when it came to the crunch, who deep-down cared about what happened to the nation's financial position when the banks were about to foreclose on you and your family was on the edge of starvation? Forget the collective: if you were clever enough to make a quid on your own and find an escape route from hard times – well, half your luck.

Contrary to what would transpire for the majority of Australians, the Great Depression was about to become a golden era for opportunists.

4

'AN OUT OF THE ORDINARY SUGGESTION...'

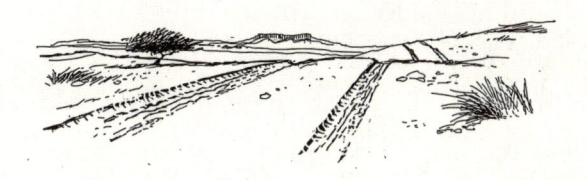

CANBERRA, OCTOBER 1929

The knockabout federal member for Kalgoorlie in Western Australia, Albert Green, had survived some extraordinary adventures in his 60 years. At various times in his life he had been a bricklayer, an outback postman, a telegraph operator and a gold miner. As a young man he left Australia for the United States and Central America, where it was rumoured he had spent time as a mercenary fighting with rebels in some Guatemalan uprising. In 1922, as the Australian Labor Party's man in Western Australia, Green won the seat of Kalgoorlie – a phenomenal undertaking, as the electorate covered practically half the Australian continent, and was the size of France, Spain, Germany, Italy, Poland and Great Britain combined. It was indeed the largest single electorate anywhere in the world – and the remote outback gold-mining town of Kalgoorlie one of the toughest and most lawless outposts in Australia. This waterless settlement only survived due to the construction of an extraordinary 300 mile-long pipeline carrying fresh water from Perth to nearby Coolgardie.

Yet for 'Texas' Green, as he was nicknamed on his return to Australia from the Americas, the month of October in 1929 was to prove perhaps his greatest challenge. Two days after being appointed as Australia's Minister for Defence on the election of James Scullin's Labor Government, the New York stock market imploded, taking Australia with it, spiralling into the greatest financial depression the world had ever seen. The irony of the US crash would not be lost on the member for Kalgoorlie – the new president of the United States, Herbert Hoover, had in fact made his own fortune on the Kalgoorlie goldfields back in the 1890s, and like Texas Green was now struggling for his political life. The implications of the crash for Australia were staggering – the country was as good as bankrupt. Texas Green and the rest of Scullin's stunned Cabinet struggled to take stock of what exactly had happened. In effect, the 48-hour-old Labor Government had been handed its death notice. Australia had no money.

Texas Green wasn't to know, but amid the melee of an election victory and the unexpected bombshell of the stock market crash, something would be delivered to his desk that would kick off a mystery that would continue for the best part of a century. Sometime in the period of limbo between his electoral triumph and his Cabinet appointment, Green received a strange, unsolicited letter of congratulation from an address in the working-class southern Sydney suburb of Kogarah. It was written in bold copperplate handwriting and was remarkably candid, offering all sorts of useful ideas and advice for the new government. And there was a revelation within the letter that opened Texas Green's eyes as wide as saucers.

Hon. Texas Green, M.H.R.,　　　　　　　Orient Road,
<u>CANBERRA.</u>　　　　　　　　　　　　Kogarah, N.S.W.
　　　　　　　　　　　　　　　　　　　14th October, 1929.

Dear Sir,

I venture to preface this letter by fraternal congratulations on Labor's Victory and to place before you a somewhat 'out of the ordinary' suggestion in regard to developing the mining industry, and incidentally enabling agricultural and pastoral industries to follow. For the past 18 years I have known of a vast gold bearing reef in Central Australia, but it is absolutely useless without water. Assays taken over 14 miles of the reef show values of 3 ounces to the ton. The reef would carry 1000 head of stampers if water were available. There is no doubt that £5,000,000 capital could be found in England to work this mine were the water difficulty overcome.

My suggestion is this:- That a flying survey be run from the waters of the Gascoyne river across W.A. with the prospect of finding a practicable line for piping water. I believe that a dam could be built in the Gascoyne of sufficient elevation to permit of the water gravitating inland. I am a competent Surveyor and prospector and would run the survey for £2,000, with sufficient accuracy and attention to detail to enable a correct estimate of its possibility to be arrived at. There is also the possibility of other fields of a payable nature being found along the 800 miles to be traversed. I would respectfully suggest that the Federal and W.A. Governments combine and share the expense to the amount mentioned (£2,000) while English capital is keenly interested in mining projects in W.A. and the Northern Territory.

I am sending a copy of this to the State Minister for Mines in W.A. hoping that action will be taken for mutual benefit.

　　Yours faithfully,

　　(Signed) L.H.B. Lasseter.

Green wasn't sure what to make of the letter. *A vast gold bearing reef? Fourteen miles of it!* The experienced Western Australian gold miner had never heard of a find remotely like it. Surely it couldn't be. Yet it would no doubt have crossed Green's mind that, as ludicrous as the story sounded, there could be a modicum of truth to this extraordinary tale. Incredibly, strange discoveries like this did on occasion actually happen. In Green's lifetime, there had been a gold-finding experience that had placed Western Australia squarely on the world map. In 1893, three Irish adventurers, Paddy Hannan, Tom Flanagan and Dan O'Shea, were following a string of gold prospectors to Mount Youle in Western Australia when one of their horses threw a shoe. While tending to the horse the three concluded that they had found signs of gold in the area, and decided to stay put. On 17 June 1893, Hannan staked a claim, which sparked a gold-fuelled stampede, pouring thousands of men into Western Australia from around the world. Kalgoorlie – a town initially called Hannan's – was born.

But this wild claim from some unknown bloke in a Sydney suburb seemed unlikely. Yet, on the face of it, as far as Green was concerned, this Lasseter fellow who had written in with his 'out of the ordinary suggestion' seemed to know what he was talking about. His letter – articulate and confident – could very well hold the secret to bailing Australia out of the global economic depression. Lasseter's correspondence wasn't simply a rant about his supposed find; he also touched on well-worn points that had been hotly discussed in parliament and in the front bar of pubs nationwide for some time. Green was well used to having his ear bashed regarding the use of all that 'wasteland' out in the middle of Australia; of the need to somehow harness all that wet-season water flooding south for irrigation and to turn turbines, thus creating a giant food bowl. But the combination of Lasseter's fantastic, out-of-the-blue gold discovery and his adjunct construction proposal for propagating agriculture in the wilds of the outback might just be feasible.

Building a pipeline from the Gascoyne River at Carnarvon in north-western Australia to the centre of the continent sounded ambitious at the very least. However, there was a precedent, in the comparable construction of the water pipeline from Perth to the Coolgardie goldfields, which had been completed after employing tremendous manpower. Big infrastructure projects like this would be needed to stimulate employment. But more than that: imagine the spectacle of a battery of a thousand head of gold stampers in operation; imagine the sound of the mighty mechanical hammers smashing quartz apart; imagine the money generated from harnessing such a phenomenal resource – a 14-mile-long block of gold.

Green chose to temper his enthusiasm, remaining circumspect on Lasseter's revelation, fully aware that previous governments had sometimes fallen for wild, unsolicited developmental schemes and ultimately come unstuck. Instead the Minister for Defence and member for Kalgoorlie opted to seek a second, more considered expert opinion.

•

In finding an expert to assess this out-of-the-blue proposal, it was as though the planets had aligned for Texas Green – for it is doubtful there was anyone who was more experienced to pass comment on this Lasseter fellow's unusual claim than the 'Consultant on Development to the Commonwealth Government', Herbert William Gepp.

Gepp was quite possibly the most qualified person of his era to report on the complex scientific and logistical requirements in mounting a mining operation like this in the far-flung corners of the outback. Aged 52, Gepp was something of a mining/business/

scientific/agricultural/metallurgical polymath. His lifelong passion for learning and unstoppable work ethic combined with his scientific abilities and business acumen made him a force to be reckoned with in both the public and private sectors. Precisely the backup Texas Green needed – he was more of a 'big picture' man.

Gepp began work as a chemist with the Australian Explosives and Chemical Company in Melbourne before taking a position with Nobel's Explosives in Glasgow. His return to Australia saw his career take off – Prime Minister Billy Hughes seconded him from the Australian Imperial Forces during World War I to investigate the manufacturing of munitions in the United States. He became renowned for applying scientific methods to economic and political problems in the interests of development and efficiency. Metallurgy might have been his calling, but his true talents lay in solving problems within the mining industry – and appointed as President of the Australian Institute of Mining and Metallurgy in 1924, his credentials were impeccable.

An in-demand public speaker, Gepp possessed a kind of personal magnetism, and audiences found his enthusiastic personality infectious. He was described as being inspirational, having 'a rare and splendid gift – the gift of being able to lift people out of mundane ruts, to set in motion new directions and habits of thought, and to excite them to attempt things of which they would probably have never thought themselves capable. He was able to raise the sights of able, but otherwise ordinary, people to the contemplation of the mountain peaks.' Quoted as saying he was always prepared to give an idea 'a run for its money', clearly Gepp was a man of vision – and from the letter sent to Texas Green he was now envisaging that there was something out in the western Northern Territory worth investigating. Yet, with the nation in the grip of an economic depression, Gepp's report to the Prime Minister's Office on Lasseter's claim ultimately recommended that the government show caution in putting money into such a scheme – but Gepp's boyish, raw enthusiasm shone through. This Lasseter fellow possessed a fantastic story, and to try to prove it would obviously be a gamble. But Australians love a gamble . . .

Naturally, Gepp's first priority in putting together his report to government was to organise a face-to-face meeting with the letter's author himself – something which correspondence held in the National Archive suggests was no easy feat. For whatever reason, the mysterious Harold Lasseter was hard to pin down. Letters between Gepp and Lasseter reveal that several appointments in Sydney between the two were missed or cancelled and then rescheduled, the federal government-appointed inquisitor ultimately having to work around Lasseter's work hours at a pottery works in the inner city suburb of Redfern. Finally it happened. 'I arranged for Mr Lasseter to meet me in Sydney on the 14th November,' wrote Gepp in his report to the minister assisting the prime minister, 'and I was fortunate to obtain the assistance at this interview of Dr L Keith Ward, my colleague on the Royal Commission on the Coal Industry and Director of Mines of South Australia. Dr Ward has considerable knowledge of Central Australia, and has reported from time to time on various matters not only for his own government, but for the Federal Government.'

Gepp then outlined the crux of Lasseter's fantastic story. 'Mr. Lasseter advises that eighteen years ago he traversed an area about 250 miles west south-west from Alice Springs, and near the western end of the MacDonnell [sic] Ranges and came across a quartz-ironstone outcrop about 14 miles long, from which he gathered samples, principally from "floaters". Mr. Lasseter carried a considerable quantity of this sample until his horse died near Lake Amadeus, and he then broke down the sample to about 5 lbs in weight, and carried it himself. The sample was analysed and returns showed about 3 ozs gold per ton.'

Gepp continued, outlining that he and Dr Ward would like to recommend at the next geological conference in Brisbane that 'the Commonwealth Government might consider the advisability of equipping prospecting parties under scientific guidance to investigate the mineral occurrences in the older rocks of Central Australia where these are exposed'. As far-fetched as Lasseter's story might have seemed, the allure of what might lie out there seemed to have bitten Gepp too. It was as though Gepp – while wanting to appear cautious and circumspect in his report to the minister – couldn't wait to put together an expedition to head out to the badlands and find out if what Lasseter said was true. The story taking shape in Gepp's mind was bringing all his fields of expertise into play: geology, science, business, logistics, technology and adventure. 'Our idea,' Gepp's report expounded, '. . . was that properly equipped parties, supplied with motor transport of the six-wheeled type, and fed from time to time, if necessary, by an aeroplane service, would be able to make a much closer examination of the mineral possibilities of the centre of the continent than has so far been undertaken.

'Dr. Ward agrees that this particular area is well worth detailed prospecting for minerals, but from the above statement of Mr. Lasseter's, it will be seen that the only definite evidence is that there is gold bearing stone in the area, of unknown extent, either in length or width. Mr. Lasseter states that the lode is very wide, but this does not, of course, prove that the whole lode is auriferous [gold bearing]. Again the gold content of "floaters" does not necessarily indicate the gold content of the reef in bulk.'

Whatever the content of the goldfield Lasseter claimed to have discovered was for the moment irrelevant, as no-one knew its location and the only person who knew of its whereabouts was in fact the first stumbling block. For the moment, the reef's secret location would remain just that. 'Mr. Lasseter maintained to us that he was unwilling to disclose the exact location of the area unless the Government would first of all provide a water supply so that if there were a gold rush, men would not die of starvation and thirst along the route. Mr. Lasseter's proposition is to make a survey of approximately 800 miles between the headwaters of the Gascoyne River in Western Australia, and he estimates that the cost of a wooden stave pipe line would be approximately £5,000,000, but this probably does not include pumping stations at intervals if necessary, and in any case is a rough figure supplied by the Wood Pipe Company to Mr. Lasseter.' Gepp conveniently refrained from commenting on the wisdom of hammering wooden stakes into termite-infested outback soil, but he continued nevertheless.

'We emphasised to Mr. Lasseter that it would be impossible to give consideration to any survey for water supply or any other matters of this sort, pending the most thorough examination of the area to determine whether there was gold in quantity and of a payable nature. We told Mr. Lasseter that we could do nothing more than report the result of our conversation and advise the Commonwealth Government of our joint views so as to provide the Government with all necessary information to enable a decision to be made.' Yet, having said that, Gepp brushed aside his previous caveat as a mere formality. He listed a five-point summation – where his first recommendation was a showstopper. 'We think,' he wrote:

1 That the area mentioned by Mr Lasseter is worthy of full investigation.
2 No detailed evidence is available of the size and value of the occurrence reported by Mr Lasseter: and he may find some difficulty in locating the spot again after the lapse of so many years.

3   There is justification for reference to the Geologists of Australia of the desirability of equipping scientifically directed survey parties to prospect certain areas in the interior of Australia for minerals.

4   If these parties were sent out, the area mentioned by Mr. Lasseter should be one of the first to be investigated. This area could be reached with six-wheeled trucks, and the party could be supplied with further food and water as might be necessary, either by trucks or aeroplane.

5   If the area proved to be <u>richly</u> auriferous, then the supplying of necessary water should first be considered from the edge of the artesian basin rather than from Western Australia. Another possibility is the damming of some ravines in the McDonnell Ranges, or the utilization of some of the supplies of water stored naturally in some of these gorges.

No doubt about it, whichever way you looked at Lasseter's tale, it all seemed too good to be true. Gepp indicated his reservations about the story to the minister, soberly pointing out there was scarce evidence 'to provide proof that there exists a profitable gold field of material size in the area reported by Mr. Lasseter, and the despatch of a party, based upon Mr. Lasseter's information, should be regarded only in the form of a gamble'. To be safe, Gepp did not recommend the Commonwealth Government put money into the expedition 'except as a portion of an organised prospecting of certain areas throughout the interior of Australia'. But he didn't think the idea should be written off as implausible – let private enterprise take the risk and the government should still be a stakeholder.

That said, Gepp then proposed an alliance between federal and state governments to act quickly in assembling the nation's most prominent geologists to discuss the idea of following up Lasseter's find. 'If you approve, we can prepare with Dr. Ward, a detailed reference for the Prime Minister or the Hon. the Minister for Home Affairs or yourself to send to the state Authority convening the Conference. This should go soon to give time for all States to prepare their story, otherwise a whole year would be lost.'

Gepp's report was remarkable. It was the work of someone clearly exhilarated by the chance to locate what would be the world's greatest gold discovery, although the imaginative five-page narrative to government was punctuated every now and then with a stern caution and a bit of obligatory concern. The report gives an interesting insight into how Gepp's mind worked – a sort of battle between optimism and caution – and could be interpreted in two ways. After perusing the dossier, a fiscally minded bureaucrat would no doubt be inclined to draw a red line through Lasseter's proposal and file it in some archive as unworkable. Yet Gepp's enthusiasm and imagination had seen him unwittingly draw up a rule-of-thumb blueprint as to how to mount an expedition to find Lasseter's gold: *properly equipped parties . . . six-wheeled trucks . . . aircraft support for reconnaissance and supply . . . costings . . .* Just to be on the safe side, the Prime Minister's Office asked for a third opinion, proposing the government geologist Dr Woolnough cast his eye over Gepp's report. Woolnough's conclusions were very much aligned with Gepp's, also cautioning against the Commonwealth tipping funds into an expedition. Nevertheless, Gepp's enthusiastic report coupled with Woolnough's conditional stamp of approval found its way across the desk of the Minister for Home Affairs, Arthur Blakeley, and then finally into the hands of Prime Minister Scullin himself, who wrote to each individual state premier stressing the urgency and importance of following up an academic discussion on the notion of a goldfield search out in the Never Never.

As Scullin recognised, the discovery of a gold reef of the proportions Lasseter was describing could help in hauling Australia out of the international economic chaos. 'In view of the economic depression existing in Australia,' Scullin wrote to the premiers, 'I need hardly say that a mineral discovery of any magnitude would at the present time be of particular national advantage.'

All this had suddenly taken life from a strange one-page letter posted by an unknown author from the Sydney suburb of Kogarah. As guest speaker at the Millions Club in Sydney that month, Herbert Gepp made a prescient observation: 'Materialism has developed a Frankenstein of trouble and is growing as the world's greatest problem . . .' As it would transpire, Gepp's report would unexpectedly sit bolt upright as a Frankenstein's monster and would soon be well and truly on the loose.

5

THE END JUSTIFIES THE MEANS...

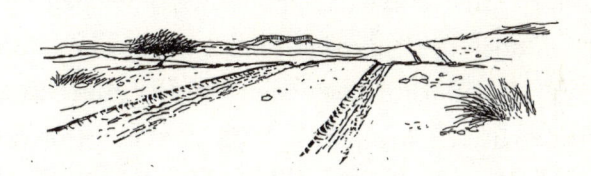

SYDNEY, MARCH 1930

If you were to suggest one example to illustrate Australia's plight during the Great Depression, you would need to look no further than the congested working waterfront surrounding the southern end of the soon-to-be-completed Sydney Harbour Bridge. The mighty single-span steel structure, simultaneously erected from both sides of the harbour, was only a few months away from being joined in the middle with giant steel locking pins. The bridge, with its enormous girders, oversize rivets and a veritable jungle of steel rigging, had gradually risen to become a symbol of twentieth-century modernity – of progress – of hope. Yet at The Rocks below, the grimy, creaking timber wharves that splayed from the harbour's edge were in stark contrast to this modernism. They were a grim reminder of Sydney's maritime trade since the arrival of the First Fleet in Port Jackson just over 140 years before.

For decades it had been an area to avoid. The outbreak of bubonic plague at the turn of the twentieth century sealed the fate of the working-class slum at Millers Point as a kind of Sydney docklands demilitarised zone. Poverty, disease, vermin, crime, prostitution, violence, filth – anyone who ventured anywhere near the area did so at great personal risk. Compounding the area's wretched reputation, Hickson Road, a long, broad avenue that ran the length of the dockside just below Millers Point's ramshackle slums, became infamously known during the Great Depression as the 'Hungry Mile'; every day, thousands of despairing workers trudged from dock to dock pleading for a job. The Hungry Mile and its long queues of forsaken men would endure as a terrible memory for many Australians – a bleak symbol of despair. It epitomised the hopeless state of the nation, also reflected in an alarming decline in moral standards. As Wendy Lowenstein wrote of the Great Depression in Australia: 'People were forced into all sorts of tricks and expediencies to survive, all sorts of shabby and humiliating compromises. In thousands and thousands of homes fathers deserted the family and went on the track (became itinerant workers), or perhaps took to drink. Grown sons sat in the kitchen day after day, playing cards, studying the horses and trying to scrounge enough for a threepenny bet, or engaged in petty crime . . .'

For opportunists who had chosen to operate outside the rules, the misery the Great Depression inflicted on thousands of Australians allowed for certain situations that – when handled carefully – could be exploited. Those in positions of influence and power were quite capable of recognising opportunities and seizing the benefits when they arose. And it didn't matter how – the end justified the means.

•

In March 1930, the Sydney office of the Australian Workers Union (AWU) within McDonnell House, 321 Pitt Street, was awash with officials and union members coming and going almost in a frenzy. The effects of the newly arrived Depression had hit hard, and if finding work was paramount, it was an almost impossible task. Long before skyscrapers appeared on the city's skyline the AWU building stood as a particularly grand Italianate structure, a dizzying eight storeys high. Yet within its walls, it was as though the place was on a war footing. The union was waging simultaneous industrial campaigns for shearers and miners, and only three months earlier a lockout at a colliery in the New South Wales town of Rothbury had seen one miner shot dead and 45 injured in a riot with police. The *Daily Telegraph* described the skirmish as 'the most dramatic industrial clash that has ever shocked Australia'.

Times were tough and violent, and those at the top were considered much the same.

It was during a court of inquiry back in 1923 that a witness had expressed his fears about AWU boss John 'Jack' Bailey's abuse of power, saying he was 'afraid of any committee on which Mr Bailey was the dominating influence'. At the time, Bailey was under investigation for tampering with delegates' votes during a national Labor Party conference: he had employed a carpenter to construct timber ballot boxes with a secret sliding panel through which he could add or withdraw ballot papers. The boxes themselves were quite an elaborate affair, each sliding panel camouflaged with bogus nail heads as though they were rock solid. 'Ballot Box' Bailey, as he was to become known, was found guilty and expelled from the Labor Party, but such was his influence that he was quietly restored to the party within a few years. He sought to clear his name so that he could run for a parliamentary seat, demanding another investigation into the affair, in which he was exonerated by the AWU; unfortunately for him this tipped into a third, independent inquiry, where he was found to be guilty after all.

Bailey was a force to be reckoned with. Those who engaged with him did so at their own risk, the former shearer being described as a 'bare knuckles fighter who literally fought his way upwards'. His arch enemy, New South Wales Premier Jack Lang, himself a political street-fighter, described Bailey as 'quick of temper and rough of tongue' and said that he 'aimed to be the ruthless party boss'.

The Great Depression had cemented Bailey, as head of the AWU, as one of Sydney's most powerful and dangerous figures – something an investigation within his own party made startlingly clear. In short, the investigative committee saw him as a crook, a thug and a liar. Their character analysis of Jack Bailey is startling: 'We do not pretend to have given an exhaustive account of Mr Bailey's misdeeds, or to have fathomed the depths of his infamy, but the outline here presented discloses an unparalleled career of criminal conduct and diabolical plotting to ruin adversaries, and of Mr Bailey's readiness to swear to the most outrageous falsities and to procure others to do the same. The cumulative effect of the evidence has forced the committee to express in definite terms its opinion that Mr. Bailey is a political leper, whose participation has polluted and corrupted every movement with which he has been associated, and without hesitation we give it as our finding that in public life he has been a menace, not only to the Labor Party and the union to which he belongs, but also to the body politic.'

Bailey's account of a meeting with a mysterious fellow named Lasseter one day in March 1930 is contained within the files of the Mitchell Library – an account written shortly before Bailey's death in 1947, some 17 years later. Its purpose was to set the record straight for anyone wanting to know Bailey's account of his role in a mystery that would last for generations. Simply, there was Jack Bailey's version of events, and then there were just flat-out lies. As he forcefully put it in his 13-page account titled 'The History of Lasseter's Reef', 'I write these comments for the benefit of students to protect them from bounding publicity agents that have not a scintilla of truth in their publicity'. As it would transpire, Bailey would not be the only one to present to the world their take on the catastrophe that unfolded in the Australian wilds during 1930–31, but it was in his office where the excitement, promise and altruism exuding from Herbert Gepp's report was seized, manipulated and slowly strangled.

6

GOLD

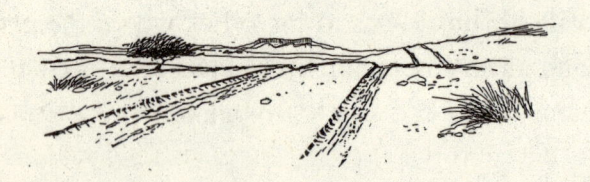

On a miserable, rainy day sometime in March 1930 a short, thickset, nugget of a man pushed his way into the foyer of the AWU office in Pitt Street and barrelled up to the woman manning the reception office. He possessed the look and build of a sideshow canvas-tent wrestler or a strong man you might see at the Royal Easter Show. His body filled out a drab three-piece suit, the chain from his pocket fob watch slung across his vest. There was something about him that exuded confidence, and to the woman at the desk his rugged, broad, weather-beaten face suggested he could be just another of the countless union delegates who frequented the building, who had sprung into the industrial workplace world from the depths of a coal mine or a backblocks shearing shed. He announced, in a surprisingly articulate manner, that his name was Lasseter and he was there to see the union president, Mr John Bailey.

Lasseter was escorted upstairs and through the labyrinthine corridors of McDonnell House, and eventually ushered into the union boss's office, where he found Bailey waiting with three other eager parties: his son Ernest; John Jenkins, the union's mining representative; and a noticeably fidgety newspaper reporter, Errol Coote. They all shook hands and seated themselves in preparation for a serious discussion. Lasseter had a story to tell.

In years to come, John Bailey would claim that Lasseter had arrived inexplicably at his office, unannounced and unsolicited. Whatever Bailey made out in his statutory declaration now held in the Mitchell Library, it is quite possible, perhaps even likely, that Lasseter was there at the AWU president's invitation.

Some months earlier, Herbert Gepp's report on Lasseter's gold reef had landed on the desk of the federal Minister for Home Affairs, Arthur Blakeley, whose ministerial role incorporated the administration of Central Australia and the Northern Territory. Blakeley acknowledged Gepp's suggestion for the Commonwealth to refrain from funding full-scale prospecting expeditions, fundamentally kyboshing requests, but clearly the minister was yet another who had been bitten by Lasseter's incredible tale of miles of gold lying untouched in the outback. Intriguingly, Blakeley the new federal minister and Bailey the union chief had more than a few interests in common. Blakeley, like Bailey, had been a vociferous Labor anti-conscriptionist during World War I. Also like Bailey, he had been a bare-knuckle union representative for shearers. But even more fortuitously, perhaps coincidentally, Blakeley had also served in precisely the same role as Bailey, as president of the AWU. As it would turn out, Blakeley's interests in the matter would become gradually more transparent. It seemed everyone wanted to be in on the Lasseter ride.

•

It's not difficult to imagine the degree of intrigue mounting within the confines of Bailey's office that day. Just who was this strange fellow Lasseter they were dealing with – and what exactly was he about to say? As far as Bailey was concerned, all he knew was that by the end of the hearing he would either be weighing into Lasseter's story boots and all or he would have this time waster thrown out into Pitt Street.

The four-person reception committee eyed Lasseter with great interest, as he glared straight back at them with an unmistakable air of suspicion. Bailey invited Lasseter to begin the conversation and the four listeners waited anxiously.

'It was a long time ago, thirty years . . . back in the late 1890s . . .' Lasseter recalled wistfully to his audience the time when, as a young man working on coastal luggers around Cairns in the far north, he decided he would give the seafaring game away and head off on horseback to cross Australia to try his luck on the goldfields of Kalgoorlie.

He left the ship, bought a couple of horses, loaded them with provisions and set off. Along the way he intended to try his hand fossicking for rubies in Central Australia. He had heard there was a ruby rush in the McDonnell Ranges west of Alice Springs, but he discovered the supposed 'rubies' were nothing more than garnets – ruby-coloured crystalline quartz. Lasseter explained that the maps in those days indicated that the McDonnell Ranges extended much further west, and being a young fellow with not much experience as a bushman, he eventually ended up in trouble – a lot of trouble. He had gone off course somewhere out west of the ranges and had become hopelessly lost. Things were looking grim. He had already had one horse die and the other was very soon about to 'do a perish', and Lasseter himself wouldn't last much longer, as he had just about reached the last of his rations.

It was then, in the middle of this dire predicament, that he stumbled onto something. He noticed a heap of small stones strung out in a line, as if they had been dumped there for road making. Their strange colour caught his attention, and so he picked up a couple and broke them. 'To my astonishment I could see fine flakes of gold,' said Lasseter. He knew it sounded incredible, almost impossible to believe, but his was a discovery unlike anything you could possibly imagine – a vast quartz outcrop spectacularly strewn with gold that stretched as far as the eye could see.

Gold. And more gold. And even more gold.

Lasseter had discovered a reef of glistening gold unparalleled by any find in history – an El Dorado that by his reckoning was about 12 feet wide, four to seven feet high and stretched for more than seven miles into the distance. 'Everywhere I examined it I saw gold,' he expounded. 'Excitedly, I followed the outcrop

– that's what it was – for some distance. Then it disappeared. But soon I picked it up again, and climbing a rise I could see it extended for miles.'

Despite being in such a terrible condition, Lasseter assured everyone he was no fool. He knew the importance of noting specific landmarks in order to relocate the reef should he somehow survive; with this done, he then set about filling an empty oatmeal bag with whatever samples he could carry.

When he left the reef behind, things only grew worse. Lasseter decided to push further west in hope of stumbling onto some far-flung homestead or outpost. For days he staggered and stumbled deliriously across countless sandhills and salt lakes – no food, no water – eventually collapsing, about to succumb to the merciless grip of Australia's dead heart, now squeezing the last vestiges of life from him. But it was not to be, as he explained to his four listeners, now hanging off his every word. As incredible as it sounded, Lasseter was granted an even greater stroke of luck. While he was lying unconscious somewhere near the border of the Northern Territory and Western Australia, a lone Afghan cameleer happened upon his sun-bleached, near-dead body, his vice-like hand still clutching the oatmeal bag of gold specimens. The cameleer took him to a government surveyor's camp, which as it happened was fortuitously nearby. On death's door, Lasseter was taken in by a surveyor – a chap named Harding – who gradually brought him back to sufficient health, and in time the pair travelled to Carnarvon on the Western Australian coast. Harding believed the gold specimens Lasseter had shown him were the best he had ever seen and he badgered the prospector to take him back out to the wilds to try to find the reef. But Lasseter's ordeal way out in the Never Never had such a marked effect on him – he was 'haunted by the nightmare of my dreadful experience' – that he was unwilling to return. Harding remained in

Lewis Hubert Lasseter, who claimed he'd discovered a vast reef of gold somewhere west of Alice Springs. Intelligent and complex, in 1924 he changed his name to Harold Bell Lasseter after American author Harold Bell Wright, who wrote a popular novel about a lost canyon of gold. (*Author's collection*)

Moving to the United States in 1901, Lasseter married and became a practising Mormon. While in America he obtained certificates in surveying, mapping and agriculture through a correspondence school – it has been remarked they could be signed in Lasseter's own distinctive handwriting. (*Mitchell Library, State Library of NSW*)

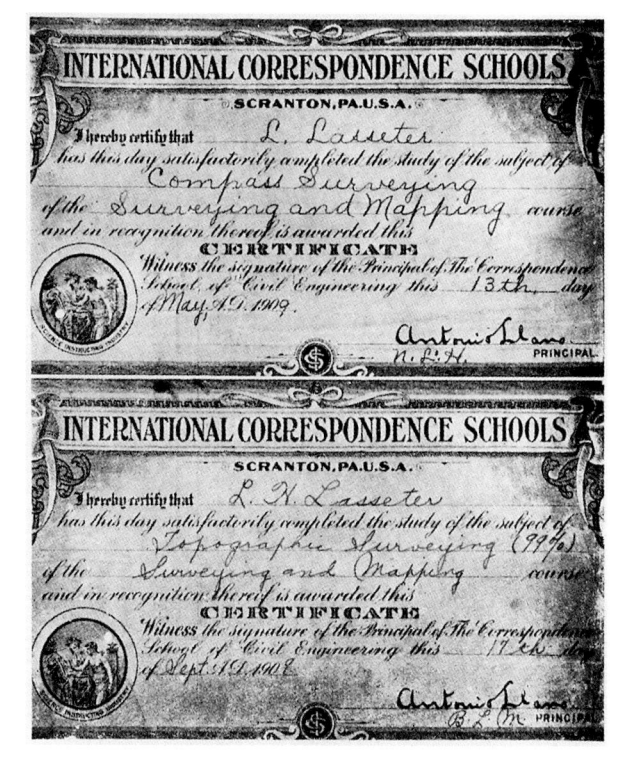

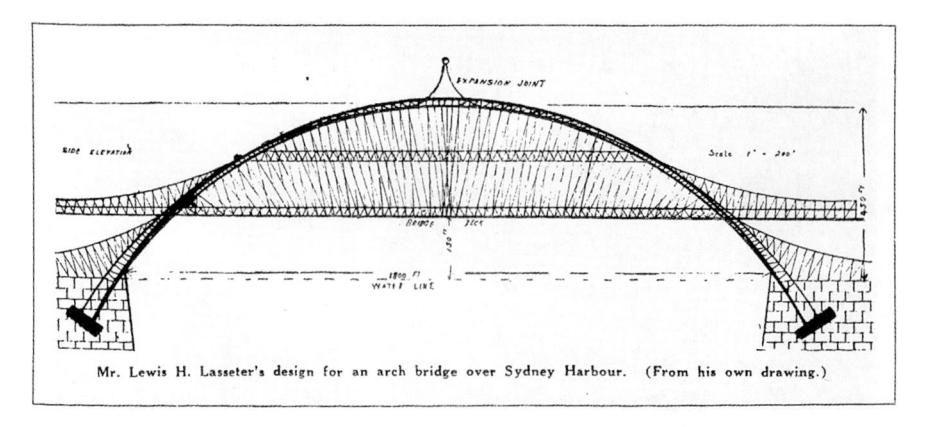

Mr. Lewis H. Lasseter's design for an arch bridge over Sydney Harbour. (From his own drawing.)

In 1913 Lasseter submitted this design for a single span bridge for Sydney Harbour which was rejected. He forever maintained Chief Engineer John Bradfield stole the idea. In 1929 he invoiced for six months labour spent on the design. (*Mitchell Library, State Library of NSW*)

Australian Workers Union Headquarters, McDonnell House, 321 Pitt Street, Sydney. It was here in March 1930 that Harold Lasseter presented to union president John Bailey his proposal to rediscover his gold reef. It was something of a Faustian pact. (*Author's collection*)

Lasseter proposed numerous grand ideas during World War I including taking 50,000 men to dig a Panama Canal-style trench across the Gallipoli peninsula. Lasseter tried several times to enlist in the AIF and was eventually discharged for medical reasons. His examiner's report is startling. (*Author's collection*)

Members of the Central Australian Gold Exploration Company expedition – claimed to be the best prepared mining exploration party ever seen in Australia. The expedition members, however, were less than prepared. (*From left*) Pilot Errol Coote, miner George Sutherland, driver Fred Colson, leader Fred Blakeley, mechanic Phillip Taylor and guide Harold Lasseter at front, with hat. (*National Library of Australia/Chrome Collection*)

While preparing for departure in Alice Springs, expedition members became suddenly aware of Lasseter's idiosyncrasies. Even during this group photo he continued to fuss compulsively over his belongings – Errol Coote described him as being 'infernally busy'. (*Mitchell Library, State Library of NSW*)

Australian brochure for the British-made Thornycroft A3 truck loaned to the expedition. Built to War Office specifications for 'trackless desert exploration', the expedition's success was largely – and unwisely – pinned on the truck's reputation for invincibility. (*National Archives of Australia*)

Expedition leader Fred Blakeley. He was a particularly competent bushman and brother of federal minister Arthur Blakeley. His authority was white-anted from the very start. (*Mitchell Library, State Library of NSW*)

Expedition pilot Errol Coote alongside the cumbersome and problematic Thornycroft, of which he was no admirer. 'We cursed that truck' wrote Coote. (*Mitchell Library, State Library of NSW*)

The ferocious drought turned creek beds into treacherous, dust-filled traps – the vehicles continually bogged. Here the Thornycroft and Fred Colson's Chevrolet truck negotiate a crossing using coconut-fibre matting. (*Mitchell Library, State Library of NSW*)

The fortress-like resistance of the mulga scrub was something no-one had expected on the drive to Ilpbilla. The expedition members would chop a path through the hard-as-iron trees for the vehicles to inch forward.
(*Mitchell Library, State Library of NSW*)

With patience wearing thin it was decided to use the Thornycroft 'like an army tank' and smash it through the trees. It worked for a while – but eventually the machine came off second best.
(*Mitchell Library, State Library of NSW*)

constant contact with Lasseter, who went on to do itinerant work around the Western Australian goldfields – badgering him to return to the site, and after three years of coaxing and cajoling, Lasseter reluctantly agreed to take him there.

The pair boarded a lugger and sailed up the Western Australian coast, back to Carnarvon, where, with a string of camels in tow, they headed due east.

Lasseter struck it lucky again. Errol Coote later recalled Lasseter's words: 'I remembered the landmarks, and we found the reef again. All the horrors of the last trip had come vividly before me, but the relocation of the El Dorado completely compensated for that now.'

Lasseter and Harding traced the outcrop for 10 miles or more, taking samples along the way, and once back in civilisation the specimens were assayed at three ounces to the ton of high-grade ore.

But the pair had made a miscalculation. After returning to the coast, they realised that their watches were an hour and 15 minutes out – meaning the bearings they had taken with Harding's sextant to record the reef's location were incorrect too. Bailey looked around the room at his companions. This miscalculation meant Lasseter's find could be anywhere.

'But this didn't worry us,' chirped Lasseter reassuringly, 'as I knew I could easily find it by the landmarks.'

Lasseter described how Harding had tried to raise the capital to form a company to go back and commence mining. But as he explained, the Kalgoorlie goldfields were booming, and 'mining men turned a deaf ear to us'. It seemed the reef's location was too remote to justify setting up a full-scale mining operation when gold was being ripped out of the ground in Kalgoorlie. Harding continued to hunt for backing; he 'hawked our show to Adelaide and Melbourne but could not find anyone to give financial support'. Harding had even sailed to London but still could find no backer. 'We drifted apart and then Harding died...'

Coote asked the question. 'Whereabouts, approximately, is the reef?'

'Near the border of Western and Central Australia,' Lasseter explained. 'I cannot tell you exactly where, but it is in either Central Australia or just over the border in Western Australia. Had our instruments been correct I could give the exact bearings, but I will have to depend on landmarks. I found it by landmarks with Harding, so there's no doubt I could do it again.'

Bailey remained calm, still absorbing the tale he had just heard. 'All right, I'll see what can be done.' He nodded to Lasseter. 'We will help you anyway. The discovery of a reef like that would wipe out the Depression overnight. It's just what Australia needs. Look in, if you are passing, in a couple of days' time, and I'll let you know what happened.'

Lasseter stood up, thanked the four for their time and wished them good day. The room was filled with an anxious, palpable silence as they watched their strange visitor pick up his hat and shuffle through the door.

Bailey turned to his compatriots. 'Damn me if we don't give it a go,' he exclaimed. The AWU boss had made his decision. Now he was simply lining up his ducks. 'We'll check up on it first. And to hell with Federal Government. They would grab everything anyway and we would get nothing.' For all Bailey's philanthropic bravado as the workers' representative, he didn't want some meddling government – and implicitly the Australian people – getting their grubby hands on perhaps the greatest gold strike in history.

Coote was the first to ask the question on everyone's lips: 'What would a reef like that be worth?'

John Jenkins, the union mining secretary, ran his fingers through his hair, his mind reeling, doing the calculations. 'If she runs for ten miles,' Jenkins began, 'is 12 feet wide, and averages three ounces to the ton down to only 100 feet we could write cheques for millions at a time.' He picked up a pencil and began scrawling a few sums, puffing on his cigar – 'like a steam engine', recalled Coote. Jenkins looked up from his arithmetic, his face frozen in shock. 'She would produce gold to the value of £66,000,000.'

Coote later recalled that precise moment; of the sudden, staggering realisation of the extraordinary opportunity presented to them. 'I stood there with my mouth open. Young Bailey tugged at his collar as if it were choking him . . . John Bailey sat like a sphinx staring at the wall.'

Lasseter's story had now taken the four of them prisoner, and despite the uneasy spectre of niggling implausibilities, no-one had any intention of trying to escape. 'The spell of gold was on us,' Coote recalled. 'Wild flights of fancy were galloping through our brains. We were already digging it up by the shovelful. Utter surrender to the God of Gold was now plainly manifest.'

•

Astounded by what had just taken place in John Bailey's office, Errol Coote and the young Ern Bailey left McDonnell House, finding a table in the winter garden at the Australia Hotel where they could discuss the matter further. Coote, as it transpired, had many interesting sides. Apart from being a newspaper reporter, he was an amateur pilot and saw an opportunity to cement himself in the role as the expedition's aviator. He was indeed putting the cart before the horse: qualified as a pilot he might have been, but licensed he wasn't. His permit had expired more than a year before. But for Coote, this was merely a trivial, technicality. He later recalled putting forward a proposition to Bailey's son: 'I've got an idea this job calls for an aircraft.'

Over the next few hours, Ern Bailey's and Coote's pub-fuelled imaginations ran wild, the conversation becoming more excited with every word. Somehow, they imagined, they would even acquire a Charles Kingsford Smith-style Fokker Trimotor. 'What a welcome we'd get landing back at Mascot with a cargo of bullion on board,' exclaimed Coote. 'We'd get a bigger reception than if we had flown the Pacific. The crowd would go mad!'

Young Bailey went even further: 'And what about keeping a small bag of nuggets to throw among them. Gee! They'd go crazy. Cripes, we'd be mobbed with cash. London would chip in too; and we would have millions behind us.'

Coote surreptitiously then asked Bailey's son how they would fund such an enterprise. Ernest already had the answer worked out. 'Well,' he said, 'the old man has a fair circle of people who would put cash into the show on his say-so, and we should arrange things so that a large number of people come in with small amounts, say about 20 each. We should limit the number of shares to about 40, and make the minimum 20. If 10 of us hold a meeting, and bring along 10 others, then we can arrange for a bigger meeting, where everybody who was at the first meeting can bring on 10 more. A sort of chain business.'

7

THE BICYCLE BUSHMAN

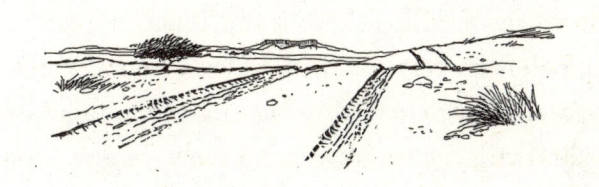

The kernel of an idea for an expedition to the wilds of the Northern Territory had been planted, yet in 1930 the region was probably the last place on the continent anyone would consider visiting. It was a land never given any consideration. Even the boundaries of the Territory itself had been arrived at by default. The states of Western Australia, South Australia and Queensland had defined their outer borders in 1862, and what was left over inadvertently defined the Northern Territory. A year earlier, when the northern boundary of South Australia was simply drawn across the middle of the continent, no white man had ever set foot in the region, let alone had any understanding of it. It was mysterious and unknown – no-one had the vaguest idea of what lay out there in the middle of Australia. Deserts perhaps? Mountains? Rivers? Maybe the vast inland sea many suspected? What there was, in fact, was almost 520,000 square miles of wilderness, almost all of it potentially completely inhospitable and potentially lethal to anyone who ventured out there. As for the issue of travelling to the Centralian portion of the Territory, in 1930 there was only a dinky single narrow-gauge railway line that poked north from Port Augusta to peter out at Alice Springs. Other than that, getting around the centre was the domain of camel drivers.

That there were considerably questionable aspects to Lasseter's story didn't seem to cloud anyone's enthusiasm. That sometime late in the previous century a very young and inexperienced Harold Bell Lasseter had set off with a few horses to walk diagonally across the Australian continent from Townsville to Kalgoorlie – the distance from Moscow to London – didn't really raise an eyebrow. And at least from Moscow to London there were roads and cities and villages with taverns and shops and shelter along the entirety of the route. People could also give directions if you were lost. Not so from Townsville to Kalgoorlie. There was nothing but scrub, desert and wilderness. That Lasseter could wander back to his reef years later, from the other side of the continent, and find it again, was also nothing short of miraculous. Indeed, Lasseter's account was riddled with miracles, but for John Bailey it didn't matter. Clearly he was besotted with Lasseter's remarkable story. It might have sounded far-fetched – but what if it was true? That was the issue.

Bailey was prepared to commit a substantial amount of AWU members' money to finance the operation and establish a company. He wasted no time in following up on Lasseter's revelation, and on the Monday caught the train to Canberra to begin making enquiries at the Department of Mines, where he turned up some remarkable information. While he found no mention of any claim by someone named Lasseter, he discovered that in the vicinity of where Lasseter had said he had been was a reef known as Harding's Reef. Further, two expeditions had attempted to locate a gold reef from the Western Australian side, but both had come under attack from natives. Men and camels were speared, expeditioners were either killed or died from disease, and those who had escaped suffered great difficulties in getting back.

Bailey unearthed a few other strange, macabre coincidences. He found a reference to a spectacular 'cave studded with copper and gold' supposedly discovered in 1895 in the Tomkinson Ranges in north-west South Australia by a prospector named Earle. Earle died 'of a malignant disease' before revealing to anyone the cave's whereabouts, and subsequent mining forays searching the ranges turned up absolutely nothing. These other parties also ran headlong into catastrophe: one expedition was attacked by Aborigines, with an Afghan camel driver speared and killed. A subsequent foray to find Earle's cave of gold had an even more gruesome result. The team's Afghan cameleer murdered one of his mining accomplices with both an axe and a .44 calibre rifle, and in a grisly reprise later returned to dig up his victim's corpse to retrieve the blanket it was buried in. The murderer absconded 300 miles into the desert, living on whatever game he shot before dying of thirst.

As gruesome as these events were, and as implausible as Lasseter's story sounded, in Bailey's mind it all seemed to be piecing together. This mysterious information about Harding's Reef and Earle's cave of gold gave just that bit more traction to Lasseter's claim.

What the AWU president rightly needed was an outside opinion about Lasseter, but not from an outsider. The Lasseter story needed to be kept 'in-house'. He needed someone who was more or less already within the Bailey camp; someone who knew and understood the lay of the land out in the wilderness and could pick holes in Lasseter's story; an experienced bushman in whom he could confide and trust, but more importantly over whom he had some control. There was one person who came to Bailey's mind, someone he had known for a long time and who had quite possibly been suggested to him for this crucial role – Fred

Blakeley, the brother of Arthur Blakeley, the Minister for Home Affairs and the recipient of Herbert Gepp's enthusiastic report recommending an investigation of Lasseter's claims.

While the idea of approaching the federal government minister's brother as an adviser might seem opportunistic – perhaps nepotistic – Fred Blakeley was indeed a sound and logical person to have on board. He had been involved in mining one way or another since he was fourteen years old, when he caught a Cobb and Co. coach to White Cliffs in north-west New South Wales to fossick for opals. He had been a prospector in the Stuart Range in South Australia, had looked for gold in Western Australia and had even searched for the elusive metal north of Alice Springs.

In his younger days, Blakeley had earned a reputation as the 'Bicycle Bushman'. He had become well known for his astonishing long-distance bicycle rides around the Australian continent. In the years before the automobile had taken hold as the dominant form of personal overland transport, Fred Blakeley was one of a hardy group of tough turn-of-the-century adventurers who had seen the bicycle as the key device for exploration of the Australian interior. In 1908, with two friends, the O'Neill brothers, Fred Blakeley cycled 2200 miles from White Cliffs to Darwin, years later recounting his adventures in his book *Hard Liberty*, described by the *Australian Dictionary of Biography* as 'one of the most remarkable books on the inland'. Like all of the 'overlanding cyclists', as this tough breed of adventurers became known, Blakeley suffered physically and mentally during these long lonely adventures, pedalling thousands of miles along bush tracks and over unmade roads, continually pushed to extremes. Overlanding was nothing less than straight-out masochism: cycling incomprehensible distances, relentlessly fighting exhaustion, thirst, malnutrition and often chronic illness. Overlanders slogged across vast deserts, through extreme heat, carried their bikes across enormous swamps and salt pans, through droughts that lasted years and floods that swirled over the horizon, and constantly dealt with the hazards of the Australian wilderness: bushfires, fetid or dry waterholes, venomous snakes and hostile Aborigines. Further, the bicycle rapidly propelled its rider into remote, uninhabited and unexplored regions, far away from civilisation and sometimes into trouble. If the cyclist experienced any serious mechanical difficulty, they could find themselves stranded beyond the reach of any rescue party.

Of necessity, these overlanders became remarkably resourceful, expert bushmen who were completely self-reliant; and as such the most successful long-distance overlanders tended to be introverts, in the correct, clinical sense – not that they were shy, rather they were energised by solitude and often found it draining to interact with others.

Certainly Frederick Blakeley could be categorised as such – a true introvert. His outback experience made him an obvious choice for involvement in Bailey's growing plan to follow up on Lasseter's story, but Blakeley's propensity for introspection and working alone rather than possessing an ability for interaction would ultimately prove a key flaw in what was to unfold.

Through Arthur Blakeley, Bailey located Fred working at an abattoir in Armidale in New South Wales and asked him to meet with Lasseter in Sydney to try to verify his claims. One Saturday afternoon the pair met. 'Some plain talking was done in Bailey's office,' Fred Blakeley recalled, 'and then Lasseter and I went into Hyde Park, where I heard his story.' And what a story it was. The pair continued to meet each Saturday afternoon for the next few weeks, Blakeley taking notes as Lasseter described his experiences from so many years before. Despite the crux of the tale being recounted more or less faithfully each time, Blakeley noticed minor discrepancies emerging. 'There were a lot of gaps in it [the story] but I felt there might be something in the yarn,' he recalled. 'Lasseter had explained the gaps by saying in reply to my questions, "If I told you that, you'd know as much as I do, and you wouldn't want me."' It seemed clear Lasseter was playing a game, and by throwing in a few red herrings along the way he would continue to hold all the cards. As Blakeley later said, these awkward conversations left him 'suspicious of Lasseter's story and of his knowledge of mining and bushcraft'. But the lure of what the tale promised overrode any niggling concerns.

John Bailey's written account describes Blakeley asking him his opinion of Lasseter and his tale. '"It's well worth investigating," I said to him. "Would you be prepared to lead an expedition to investigate it?" He said he would be delighted to do so.'

•

Blakeley's agreement to oversee an expedition was more than enough for Bailey, and he was swift to act. Bolstered by Blakeley's endorsement, he was now more than satisfied with Lasseter's story, and with an expedition leader in hand all he needed was capital to mount an expedition along the lines of that envisaged in Herbert Gepp's report. But in 1930, aircraft and six-wheeled trucks were still virtually the stuff of science fiction – prohibitively expensive and almost impossible to obtain.

Initially Bailey opted to create a small syndicate of six, based on Blakeley's somewhat naive belief (as someone whose knowledge of exploration had been accrued from a bicycle seat) that an expedition into the interior would cost only about £300 – both Blakeley and Bailey agreeing to put in £50 each. But at the syndicate's first meeting it was decided that such an amount would be wildly insufficient and that a company should be set up to raise capital of £1000, made up of shares at £20 each.

Bailey organised a meeting of more than 50 of his 'friends' – comprising chiefly union officials and some influential Sydney business identities – who gathered to hear his report on Lasseter's proposal. When Bailey theatrically unveiled his star to answer questions, the crowd jostled to see and hear the strange and mysterious prospector. This was the man who had discovered the greatest gold find in history. He had been out there, and he had seen it, he had touched it. He'd had it in his hands.

Lasseter again related his story, mesmerising almost all who listened. For a solid hour he was questioned, and remained resolute and unflinching in relating his belief to a largely convinced gathering. But not everyone was so sure. As it happened, sitting in the front row was the famous Australian aviator Flight Lieutenant Charles Ulm, who had been quietly working out calculations on the back of a cigarette packet while Lasseter spoke. His facial expression suggested he was slightly dubious of Lasseter's narrative.

'How much did you say your watch was out when you reached the coast again?' Ulm asked.

'About an hour,' replied Lasseter.

'Well, as the Earth makes a complete revolution in 24 hours, it means from a longitudinal point of view, you were one-twenty-fourth of the earth's circumference out of place on your bearings. That in turn means that on your bearings the reef must be somewhere in the Indian Ocean.'

'If you take it that way, yes,' replied Lasseter, but he explained that as it was impossible to say how far their watches were out at the time they made the calculations, the bearings would be useless anyway. Besides which, Lasseter said, turning back to the crowd, he was going to rely on landmarks alone. 'They got me there before, and I fail to see why they could not get me there again.'

The room was more than satisfied, but Ulm hadn't quite finished. 'Except the fact that you have not seen those landmarks for thirty years or more,' he cautioned. 'And, although in this case, time, lends enchantment to the view, I am afraid those landmarks have become slightly hazy in the mental picture, after all this time.'

'If you had seen the country under the conditions I did,' Lasseter replied, 'you would never forget those landmarks.'

The crowd roared and, sensing the mood, Ulm responded gracefully and tactfully. 'Well, it's too hazy for me to speculate personally, but any assistance I can give to the expedition on aerial matters I shall be only too glad to do so.'

That was the attitude, and the crowd applauded. Don't waste everybody's time asking questions about Lasseter's story – just get behind it.

'The room veritably buzzed in the pause in proceedings,' remembered Coote. 'Everybody was talking gold. Eyes were bright; lips were being licked; workaday cares were forgotten. Everyone was gazing through the pearly gates of prosperity, and each man was treading along streets of gold.'

It was decided to set up a company in which shareholders could invest. Blakeley recalled how rapidly investors flocked to take part in the Lasseter phenomenon: 'From our syndicate with its £300 had come the suggested company with £1,000 invested . . .'

John Bailey took the floor. 'Well, gentlemen, we will hold another meeting tomorrow night. Bring along your friends; we want everyone to be in on this. Those who are prepared to back Lasseter, please put your names down here . . . the sooner we get on with it the sooner we will be sharing the spoils.'

'Again the audience literally "licked their chops",' recalled Coote. 'The fever had gripped them, and the company, still without a name, was an assured success.'

8

MACHINES

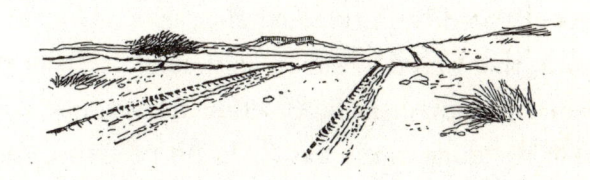

Coincidentally, at the same time as Bailey was gathering investors, a mapping expedition was travelling right through the very heart of the region Lasseter described. Under the command of the renowned professional explorer Donald Mackay, this was a bold mission championing the use of aircraft for survey and exploration work. Like Fred Blakeley, Mackay in his younger years had been a hard-bitten overlanding cyclist, in 1900 completing an 11,000-mile ride around Australia. The Mackay Aerial Survey Expedition of 1930 could be viewed as something of a prototype for the Lasseter's reef mission, the first to use a plane – in this case an Australian-built tri-motored Lasconder – for exploration in the Northern Territory's west.

The expedition would fly to Mildura, Oodnadatta, Hermannsburg, Alice Springs and then to a specially cleared landing site at a remote waterhole called Ilpbilla in the Ehrenberg Range in the far western region of the Northern Territory. Mackay's aerial survey was well planned, employing the services of a local cattleman,

Bob Buck, who had cleared a landing site at Ilpbilla two months earlier. Buck then ferried tins of fuel out to the remote airstrip using a string of 70 camels, administered by Afghan cameleers and native workers.

Prior to leaving Canberra on 24 May, Mackay and his party were feted by Prime Minister Scullin at a luncheon at Parliament House. Mackay, replying to the prime minister's farewell toast, said that 'Central Australia was known as the "dead heart"' but he believed it would be very much alive in a few years, and he was confident that rich deposits of gold would be found: 'A rich gold find in Central Australia or elsewhere would relieve our troubles as it has done in the past.' All this of course fuelled the interest in Lasseter's story.

•

It was agreed to set up a transport committee for the syndicate with Errol Coote designated chairman, but as it turned out his idea for hiring a big tri-motored Fokker aircraft was prohibitively expensive. It was decided to try to secure a smaller, less costly machine. As Coote put it, the aircraft 'would be supported by a motor truck; thus making the expedition the most modern equipped for transport that had ever gone into the interior in search of gold'. From the outset, Coote considered that organising the aircraft for the expedition was the priority. He envisaged himself as the person who would locate the reef from above. In his mind the motorised overland component crawling mile after mile on the ground was actually the support crew for his flight.

In the 1920s and 30s, aviators were feted in even grander style than movie stars, and Coote believed this would be his moment in the sun. As the expedition's official pilot, discovering the reef would put him on the map, joining the pantheon of internationally famous Australian airmen such as Charles Kingsford Smith, Bert

Hinkler and Charles Ulm. And it wasn't just about recognition. A monumental aviation achievement could prove as lucrative as winning lotto today. Only the year before, Bert Hinkler's solo flight from England to Australia had seen the penniless aviator presented with an impromptu £2000 cash prize from the federal government.

Once word was out about the expedition, sponsors fell into line, seeing the worth of being part of the great gold hunt. The NSW Government Railways promised free transport to Broken Hill, while the Atlantic Union Oil Company assured free fuel for a motor vehicle and the aircraft. McIlrath's grocery stores offered food supplies at cost price and a camping shop promised equipment at 25 per cent below cost.

With healthy sponsorship rolling in, Coote headed out to the aerodrome at Mascot in Sydney, determined to find an aircraft suitable for the expedition. He was most impressed with a supercharged Gipsy Moth biplane named the Black Hawk parked in the de Havilland Aircraft Company's hangar. It had been flown in an air race by Major de Havilland himself, posting a sensational time, and Coote, imagining himself in the cockpit, thought 'it would be the ideal bus for the job'. But when he told Charles Ulm of his find, the experienced aviator dismissed the idea of the Black Hawk immediately. 'That Racer! The Black Hawk is no good for you,' said Ulm, bringing the high-performance aircraft back to earth. Coote had in effect been looking at a hot rod when what he needed was a truck. 'In the first place it is a wooden machine and out in that country you need as much metal in your bus as possible,' explained Ulm. He was talking about the airframe – the skeleton over which the fabric 'skin' was wrapped. They needed a metal skeleton rather than a wooden one, coupled with more of a no-nonsense tractor motor rather than a Ferrari engine. As Ulm said, 'wood warps in the heat. Further a high compression job is essentially not a working outfit. She is only fit for racing and stunting. A metal Gipsy is what you need.' With a single telephone call, Ulm teed up the kind of aircraft Coote needed, and it would be available for inspection the following day. The pair had managed to tentatively secure a near-new Gipsy Moth for £600 – the impecunious Coote immediately notified Ern Bailey, who then organised £100 deposit. Ulm flight-tested the machine and gave it the thumbs-up, pronouncing it 'excellent'. The deal was made and Coote named the biplane the *Golden Quest*. 'It looked striking,' enthused Coote, 'with its black wings, red fuselage and undercarriage, and gold lettering. It was just like a huge hornet.'

•

On returning to the AWU building one afternoon, Coote was informed by John Bailey that there was a fellow waiting in his office 'with a knowledge of aviation' who was interested in joining the expedition. By all accounts this chap had lived through some extraordinary adventures, informing Bailey that not only had he been the personal mechanic to Viscount Trenchard, the World War I aviator regarded as the 'Father of the RAF', but was also a former RAF pilot with the rank of flight lieutenant.

Not sure what to make of this, Coote entered Bailey's office to be greeted by a tall fair-headed fellow surrounded by large unfolded maps and newspaper cuttings. He stood up and shook Coote's hand, introducing himself as Charles Lexius-Burlington and saying that he was keen to be part of the upcoming expedition into Central Australia. Lexius-Burlington explained that he was an experienced mechanic who had been working up in the remote parts of northern South Australia for General Motors Corporation. He had also done his fair share of pegging gold leases on the Watut River in New Guinea.

Wary of another pilot attempting to muscle in, Coote coolly asked him if he still worked for General Motors, as the expedition was in need of a truck.

'No, I am not with them now,' he replied, 'and I would suggest you take a British vehicle. They specialise in trucks for those sorts of trips. In fact I think I could get the loan of a truck for nothing for you.'

John Bailey was impressed. Coote seemed less so. Lexius-Burlington had connections with the British truck manufacturer Thornycroft, a company that specialised in rugged military-style vehicles used for specialist heavy-duty work. He was on particularly good terms with the company's manager in Australia, A. G. Hebblewhite, and with one phone call confirmed that Thornycroft's representative would attend the shareholders' meeting scheduled for that evening. Just like that.

•

The rousing conference held in the union offices commenced at 5 pm, and was described by Coote as 'a bumper affair'. John Bailey conducted proceedings as if it were high theatre, building a mood, creating suspense, revealing one fabulous surprise after another, each one more exciting than the last. Bailey announced to everyone that the newly named Central Australian Gold Exploration Company, or CAGE, was now in business, preparing to send an expedition into the dead heart to locate Lasseter's reef. As self-appointed company chairman, Bailey had engineered its structure so it appeared more of a philanthropic enterprise than a money-making venture. To much spontaneous cheering and applause he announced that there would be no payment for himself, the company secretary or directors: 'All this work would be honorary.'

'What a company! What a big-hearted lot of Australians at the helm! It surely must succeed!' gushed Coote, praising Bailey's largesse. What no-one knew was that Bailey was ploughing union members' funds straight into the operation to finance it, under the guise that they were his own. But there was more.

'Now we will get down to business as soon as we have a company secretary,' Bailey declared. And who better to nominate and appoint than his own son, Ern?

The crowd duly assented, and Bailey then unveiled his expedition leader in grand style. 'In view of the fact that there is a Labor government in power in the Federal sphere, and in view of the fact that Arthur Blakeley is the Minister in control of Central Australia, I don't think we could do better than have his brother, Fred Blakeley, as leader of the expedition.'

To the shareholders, this made perfect sense; it was almost as if the plan had been government approved. Then, in sombre tones, Bailey announced that there was a gentleman in attendance who wanted to offer a contribution that would underline the strength of kinship between Britain and Australia.

A. G. Hebblewhite, the Australian manager for Thornycroft trucks, rose to his feet. His enthusiasm shone out to the crowd as he explained that he had been approached by his friend Lexius-Burlington that afternoon and had been buoyed by the opportunity for his company to be of assistance to the expedition. He was confident that Thornycroft was the 'best truck in Australia' and on behalf of the company he would be prepared to loan a six-wheeled truck for six months. 'I am impressed with the sincerity of purpose which is actuating this expedition and I am sure it will be of great benefit to the Empire as a whole and Australia in particular.'

'Loud cheering, whistling, and stamping of feet greeted this effective speech,' recalled Coote. Bailey leapt to his feet to capitalise on this marvellous tie-in of Anglo-Australian industrial relations. He began to work the crowd as he would a shedful of disgruntled shearers on a rich cocky's sheep station, winding the meeting into an even greater frenzy, urging potential investors to look on their financial input as not only part of a plan for the greater good of Australia, but also for that of England.

'Australia as part of this great Empire must bear her burdens of responsibility,' he declared before ruefully asking, 'How else can she emerge successfully from the mire of depression? What can she do to fulfil her obligations, her financial obligations, to the mother country?' The crowd murmured in agreement, for Bailey had the answer right next to him. The panacea for Australia's financial woes was five feet tall, wearing a homburg hat and sitting squarely in front of the audience. Bailey placed his hand on Lasseter's shoulder. 'Can anybody at this meeting tell us of a better way out of our difficulties than by backing this man, Lasseter?'

9

'A FELLOW NAMED BURLINGTON...'

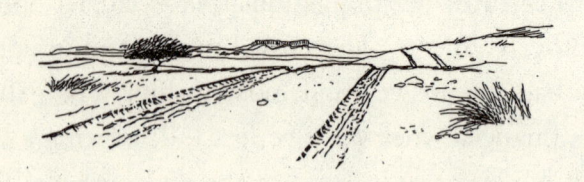

Charles Lexius-Burlington had indeed been a godsend for the expedition. Through his extraordinary connections he had secured the free use of the most vital piece of equipment for what would be one of the most arduous mining exploration ventures ever seen in Australia. Lexius-Burlington was something straight off the cover of a *Boys' Own Annual.* Handsome, tall, strapping and well educated, Lexius-Burlington combined impeccable military credentials with a record of swashbuckling adventure. A capable pilot in the RAF, he had in recent years been working with gold-mining companies in the remote regions of New Guinea, where he was a contemporary of Errol Flynn. He had written numerous serialised cliffhanger-style articles in newspapers detailing his daring exploits while employed as a volunteer machine-gunner with a punitive expedition seeking vengeance in hunting down New Guinea natives who had murdered white miners. It was rip-roaring stuff that kept readers on the edge of their seats.

Lexius-Burlington's enthusiasm for the CAGE expedition seemed inexhaustible. He suggested to Coote that they visit his friend Professor Sir Edgeworth David at Sydney University to see if he could throw some light on the auriferous potential of the region they were about to visit. The professor was convinced that there was gold in that area. 'In fact,' he told Lexius-Burlington and Coote, 'I am preparing a geological map of Australia and I show the same kind of country, extending to the borders of Western, Central and South Australia that is found around Kalgoorlie.'

Things were beginning to happen – perhaps too quickly for Errol Coote's liking.

Lexius-Burlington's achievements and connections were undoubtedly perceived by Coote as a threat to his own position as self-proclaimed deputy leader and as the expedition's official pilot. Fundamentally, Errol Coote despised him. And for a reason. As impressive as the well-connected Lexius-Burlington's service record was, Coote's own military career had been less than auspicious – his military records in the National Archives are a real eye-opener. Coote's service in the Great War was not one of gallantry or heroism and derring-do in battle; rather, he spent lengthy periods of time in military prisons for a variety of petty crimes.

Sydney-born Coote enlisted in the Australian srmy in 1917, a military path that had begun badly when he gave false details on his enlistment form, with his place of birth being cited as Vancouver, British Columbia – an invention duly picked up and rectified before his eventual discharge. Before being posted to England, Coote was charged with being absent from piquet duty – in wartime, an especially serious offence – but this was merely an entrée for a particularly unsavoury military career. Once in England, he absconded from his training camp, was arrested by the military police in London and locked up pending a court martial. He escaped again, was rearrested and escaped once more, on the day of his court martial. Things were to become even more serious. In Hawkesbury, Coote was arrested by local police – found in possession of a stolen motorbike and sidecar. Once charged, he was turned over to the less than philanthropic British military provosts, who were determined to make sure the slippery Australian petty crim would not scarper again. Further indictments were laid and another court martial was set, for December 1918. Coote pleaded guilty to his civilian charges of stealing a motorbike and sidecar and was fined five pounds or one month's gaol – and as he had already been a prisoner earning no pay, he did the time. Within the walls of Chipping Sodbury military prison he then faced his court martial, and was sentenced to four months' gaol and loss of all pay. He was released from Lewes Detention Centre in January 1919, owing the army £26 10s.

•

For the upcoming expedition, Coote decided it might be wise to reinstate his expired pilot's licence. After all, the expedition's official aviator should in all propriety have one. In a long-winded, obsequious letter to the Director of Civil Aviation, Coote asked for his 'blessing' as he had 'sadly neglected the regulations governing flying'. He ended his request for a reinstatement of his permit with a caustic aside about Lexius-Burlington, now the team's mechanic: 'A fellow named Burlington is trying to butt in on this expedition, but several Federal Members have warned us that his presence will not only be a handicap, but will be resented. He says he is a competent aircraftsman, mechanic, and flying man. Could you let me know if this is so? Anyhow he is NOT going with us.'

In any case, the Director of Civil Aviation had more pressing issues with Coote as the expedition's pilot – there were grave concerns as to whether he was proficient enough to manage what was required of him. As far as the director, Colonel Brinsmead, was concerned, even if Coote's Class A licence was reinstated, it was not a sufficient qualification for what he was proposing to do in the outback. To operate in the Never Never carrying passengers and goods Coote needed a Class B, or what we would now call a commercial pilot's licence. Brinsmead wrote to Ern Bailey expressing his concern that Coote was not up to the task, telling him, 'I cannot urge you too strongly to obtain a really competent man.'

But the ensuing correspondence from the CAGE office was ponderous and vague enough to buy Coote time. On top of that, Civil Aviation wasn't clear about the transaction of ownership during the purchase of the *Golden Quest* – precisely who owned it wasn't finalised. The aircraft was actually unregistered. An unregistered plane, with an unlicensed, amateur pilot. What could possibly go wrong?

10

'DESIROUS OF GOING TO CENTRAL AUSTRALIA...'

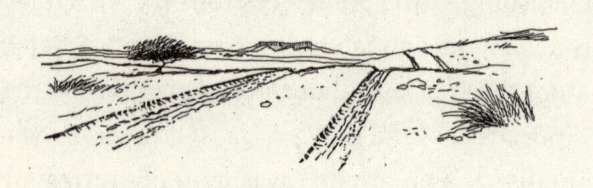

John Bailey's office had been inundated with requests from desperate hopefuls begging to be included in the forthcoming gold hunt. The thrill of it all offered a glimmer of hope and escape from a worsening depression. In the thick of these requests came one from someone 'doing a little surveying' who was 'desirous of going to Central Australia in order to complete my studies'. This mysterious individual turned out to be Captain Charles Blakiston-Houston of the 11th Royal Hussars, seconded to Australia to serve as the aide-de-camp to the Governor-General Lord Stonehaven. Blakiston-Houston's furlough was soon to commence and he was keen to see something of Central Australia before his return to England. The tall 32-year-old Irishman met with Coote at the Sydney journalist's usual haunt, the winter garden at the Australia Hotel, and outlined his experiences and qualifications. During the Great War he had seen action serving with the newly created Tank Corps, and later was a member of an (unsuccessful) expedition organised to climb Mount Everest. Blakiston-Houston had learned of the CAGE expedition through Brigadier McKay, a friend of Professor Edgeworth David, whom Coote had just met through Lexius-Burlington.

Coote was impressed – especially in that Blakiston-Houston would only be travelling partway with the expedition before he needed to head back. He was no threat to Coote, as he would be in effect nothing more than a sightseer and an extra pair of hands if needed.

For John Bailey, Blakiston-Houston's inclusion was the icing on the cake. As his son Ern agreed, 'He will be a great asset to have on the expedition. Almost gives it the hallmark of vice-royalty.' Having the Governor-General's aide on board more or less implied he was there with the good wishes of King George V's representative in Australia. And if the Governor-General – the King's representative – thought the expedition was a good idea, then therefore so did the King. And what could be better for bolstering the venture's prestige in the eyes of the outside world than having the King on board? It was a long bow – but by Jove it sounded good. As Fred Blakeley described it, the inclusion of Captain Blakiston-Houston of the 11th Royal Hussars 'gave the party a hoity toity touch'.

By now the expedition was genuinely starting to take shape as sponsors and investors fell into place. But as things grew, it seemed as if the key proponent for the ever-building juggernaut had somehow been left out of the picture. Forget federal ministers, vice-regal aides-de-camp, union bosses and high-flying impresarios with double-barrelled surnames, it was Lasseter who had ultimate control. And he decided to flex his muscles by doing something extraordinary. Out of the blue, Lasseter made it clear he did not want Lexius-Burlington on the expedition. To emphasise this point he threatened to withdraw his services unless this demand was met.

It made no sense. If you were to select – even beg – anyone to take part in the expedition it would be Lexius-Burlington. He knew his way around maps. He knew motor vehicles and aircraft – he knew how to operate them, how to service them, how to fix them. He personally knew the manufacturers. As all the shareholders had witnessed, with one phone call he had procured the use of an expensive British-made military-style truck designed specifically for arduous long-range desert campaigns, just as the CAGE expedition needed – for free. Even further, he had pegged gold claims before – and in the wild, headhunting regions of New Guinea no less. He had served time as a machine-gunner on dangerous, punitive patrols. He was on excellent terms with university luminaries, renowned geologists, business executives, military men of senior rank and those in political circles. His capacity to make things happen instantly made him tower over everyone else involved.

But perhaps that was the problem.

Perhaps Lasseter – ever secretive and evasive – saw Lexius-Burlington's no-nonsense, open-handed approach as a potential threat to his position as the operation's lynchpin. And perhaps Lexius-Burlington would be the first person to flag and act upon the holes in Lasseter's story. So far the expedition had gained traction only through Lasseter's eccentric, curious and timely handling of snippets of information. This was how he liked it, with himself as the only person holding all the cards. Nothing would challenge Lasseter's grip on the reins of his narrative.

Errol Coote was no doubt more than pleased with the removal of Lexius-Burlington. Blakeley, the expedition leader, also seemed content to acquiesce to Lasseter's demand, writing that Lasseter's 'objection had been noted' before setting about finding a replacement mechanic. Perhaps the accomplished Lexius-Burlington was perceived by Blakeley as a threat to his leadership as well.

From the outset it was clear that Blakeley didn't have a firm hold on the project that was being assembled in Sydney. The long-distance cyclist, prospector and abattoir worker might have known his way around the scrub on a pushbike, but overseeing an expensive, shareholder-funded, air-and-land-coordinated venture like this must have given Blakeley grave concerns as to whether he could actually pull it off. The last thing he needed was an accomplished, competent alpha male like Lexius-Burlington breathing down his neck. With the Thornycroft truck now secured, Lexius-Burlington was unceremoniously dropped – the CAGE expedition was a closed shop.

•

As it turned out, finding a suitable mechanic wasn't an issue. A young Englishman, Phillip Taylor, employed by de Havilland Aircraft in Melbourne, was eager to pick up some extra money, as he had recently become engaged to be married.

The team was now finalised, a small outfit comprising six men: Fred Blakeley as the expedition leader, Harold Lasseter appointed as the official guide and Errol Coote as the pilot; with them was George Sutherland, a prospector and miner; Phillip Taylor, the team's official mechanic; and, travelling part of the way as an observer, Captain Blakiston-Houston. It is important to understand these personalities in their own right – each remarkably different from one another.

For this expedition there was very good reason to assume there would be a victorious outcome. The strategy was for the party to make its foray into the interior as though embarking on a military campaign, conducted as a kind of exploration blitzkrieg – combining aircraft with a land force, maintaining communication via the latest in portable wireless technology. This technology would in effect be used to force the most remote and potentially lethal part of the continent into surrendering its treasure – a treasure that had been kept secret for millennia, until a mysterious and taciturn prospector from Kogarah happened to stumble upon it.

The CAGE expedition would largely capitalise on the work already carried out by the Mackay Aerial Survey Expedition only a few months before in setting up a temporary aerodrome at Ilpbilla waterhole in the Ehrenberg Ranges, some 240 miles (380 kilometres) west of Alice Springs. Their overland route north out from the Alice would turn west to roughly follow the Tropic of Capricorn, and after reaching Ilpbilla they would set up a base camp and then strike out for where Lasseter knew the gold reef to be. There were countless issues to be considered – not the least being the matter of Lasseter relocating his find. Perhaps the most pressing was the need to reach their objective before the onset of the fierce summer weather.

As in all successful military campaigns, mobility was the key. A dozen years earlier, World War I had delivered unprecedented development in mechanised transport. Within the space of four years, armies once completely reliant on the horse for reconnaissance, attack, and the movement of troops and materiel finished at war's end with the mass production of advanced and reliable aircraft and almost total mechanisation on the ground. And the war had had a further, unexpected side effect: a new generation of men previously oblivious and unexposed to mechanical technology, who were now capable of driving, flying and possessing the ability to understand and maintain these machines.

The days of camels and horses were over. In the twentieth century it was the internal combustion engine, benzene, electricity, flight and the knowledge that modern man could outwit the greatest impediment to bygone exploration – nature.

Of all the technology at the expedition's disposal, the six-wheeled Thornycroft truck was the item on which its success largely depended. The Thornycroft could trace its lineage to the great lumbering British War Department trucks hauling heavy artillery pieces through the mires of the Western Front, and having such a machine on the expedition no doubt gave great assurance of its success.

During and since World War I, Thornycroft trucks had earned a reputation as formidable machines – practically unstoppable, and thought of in the same way as one might think of a British Dreadnought. In fact, in 1930 the company released a model called just that. The truck delivered for the CAGE expedition was the three-ton carrying-capacity A3 model – a rigid-chassis, three-axle, six-wheeled truck built to British army specifications for 'trackless desert exploration'. Its drive train was designated six by four – that is, the truck had six wheels of which the rear four were driven. It was powered by a simple 3.6-litre side-valve petrol engine capable of producing up to 40 horsepower. The power was fed to the wheels through a four-speed non-synchromesh 'crash' gearbox coupled to an auxiliary transfer case, offering high and low range, giving the truck eight forward gears in total.

The vehicle had an interesting connection with the notion of a 'hoodoo' in the west of the Northern Territory. It was described as a 'sister truck' to the Thornycroft used to bring back the bodies of ill-fated aviators Lieutenant Keith Anderson and 'Bobby' Hitchcock who a year earlier, in March 1929, had perished after their aircraft the *Kookaburra* was forced to land in the Tanami Desert, north of where Lasseter had found his gold reef. Anderson and Hitchcock were part of an aerial search looking for Charles Kingsford Smith and Charles Ulm, who had gone missing over the Kimberley in their aircraft the *Southern Cross* during the first leg of a round-the-world air race. Unable to take off, Anderson and

Hitchcock had ultimately died of exposure and thirst. Kingsford Smith and Ulm were miraculously found at the mouth of the Glenelg River in north-western Australia. Despite the nation's relief that they had been located, certain sections of the media and the public maintained that Kingsford Smith's landing had been a publicity stunt: the incident was infamously labelled the 'Coffee Royal Affair', due to the brand of coffee and brandy the pair drank while awaiting rescue, and many believed that Kingsford Smith was responsible for Anderson's and Hitchcock's deaths. Even after an inquiry exonerated Kingsford Smith, his public reputation never fully recovered.

•

As the mighty exploration juggernaut was being readied to leave Sydney and launch into the Never Never, the company chairman John Bailey still had one difficult problem. The grand expedition, now equipped with its aircraft, a six-wheeled truck, the latest in portable wireless and surveying equipment and tons of supplies, in reality had no firm confirmation as to where it was actually headed once it left Alice Springs – somewhere near the border of Western Australia and the Northern Territory, apparently. The problem for everyone involved – the company, expedition members, sponsors, government officials, shareholders and financiers – was that the only person who held the key to where the gold reef actually lay was the secretive Harold Bell Lasseter. Not only was he holding the key but he was determined no-one would get near it.

This was a worry. Bailey was determined to nail the old prospector down on revealing the reef's location as part of a signed legal agreement. Lasseter could hold his cards close to his chest as long as he liked while in the city, but once the expedition was underway he needed to come clean. The agreement stated: 'The said Lasseter shall also as soon as reasonably may be after the expedition party has reached a point of fifty miles beyond Alice Springs disclose to each member of the proposed expedition such <u>landmarks</u> and features of country as will enable them to be of the utmost assistance to him in relocating the reef.'

But Bailey wasn't content with simply a contractual agreement. If something happened to Lasseter – if he died or disappeared – the secret of the gold reef's location would go with him. In an act that might ordinarily seem bizarre, yet was hardly surprising considering the unusual events and agreements already taking shape, an arrangement between the company and Lasseter saw him deposit a mysterious sealed envelope with the Bank of Australasia in Sydney's Martin Place, containing secret instructions as to the whereabouts of the reef. The envelope held two messages (one written in normal ink and the other in invisible ink), and was only to be opened on Lasseter's return to Sydney – or if he was pronounced dead. If such a terrible and unthinkable thing should happen, the reef's coordinates would be disclosed to the company through the office of the Public Trustee. With this secured, John Bailey could breathe a little easier.

11

PLANES, TRAINS AND AUTOMOBILES

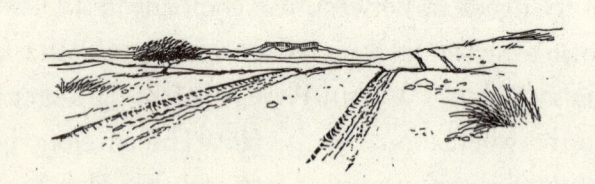

DARLING HARBOUR GOODS YARDS

A cloud of swirling smoke dispersed to reveal a hissing, coal-black 19 Class steam engine, beside which an overall-wearing guide known as a 'shunter' stepped carefully over the myriad of railway tracks and points, beckoning the driver to slowly ease the locomotive forward, joining its buffers with those on the waiting flat car. Fists of white steam shot downward from the engine's connecting rods onto the tracks, while acrid, grit-filled smoke belched from the engine's funnel. The locomotive inched closer until the unmistakable sound of steel buffers grinding together signalled it was safe to hook and chain the pair.

Lashed tightly to the flat car was the CAGE expedition's immaculate Thornycroft truck, fully laden and bound for Central Australia. Two tons of provisions had been loaded aboard in timber crates stencilled 'C.A.G.E. – Alice Springs', enough supplies to run a small town – cost-price groceries from McIlrath's stores, a full kit of blacksmith's tools, boring equipment, a water-condensing unit and an oxy-welding kit. Even though the truck was now loaded on a train ready to head out of Sydney, the logistics for transporting it by rail to the Alice were always going to be a headache. The bewildering and obstinate problem of differing railway gauges between states meant the truck would be loaded and unloaded six times. It was frustrating. The idea was for the Thornycroft to be entrained from Sydney to Broken Hill on the standard-gauge railway, from Broken Hill to Cockburn in South Australia on the 35-mile-long narrow-gauge Silverton tramway line, then reloaded onto a flatbed carriage to Quorn in South Australia, and then from Quorn to Alice Springs on the Central Australian Railway.

But from the beginning, the best laid plans went awry. With the steam whistle announcing the train's departure, the Thornycroft was freighted out of Sydney without the expedition leader travelling with it. Fred Blakeley, George Sutherland and Captain Blakiston-Houston missed the train to Broken Hill by half an hour. Hardly a good omen.

It would be Lasseter who travelled as the truck's guardian and overseer, accompanying it on the slow journey out of Sydney and across the Great Dividing Range – the New South Wales government supplying free train travel to Broken Hill, as did the city's Silverton Tramway to Cockburn – but plans fell through to send the truck on a flat car across the state. As Lasseter did not know how to drive, he hired a driver to take the truck to Quorn – the railhead for the federal government-owned narrow-gauge North-South Railway (nicknamed The Ghan, after the Afghan cameleers imported to help build the overland telegraph) to Alice Springs. Lasseter spent several long days waiting at Quorn, until the eventual arrival of Blakeley and Sutherland by passenger train. The three negotiated for the Thornycroft to be loaded onto a narrow-gauge flat car attached to an empty cattle train for the three-day rail journey to the Alice. The train they caught was the Through Mixed – part goods train, part passenger train – also known as the Dirty Ghan, as opposed to the Flash Ghan (the conventional passenger train).

•

A journey on the North-South Railway was an adventure in itself, the tracks placed squarely on the desert sand, which meant the train simply jolted its way across and over sandhills, making for particularly uncomfortable travel. (During World War II, Australian soldiers travelling along the spectacularly bumpy train line nicknamed it 'The Spirit of Protest'.) The Through Mixed took much longer than the passenger train to reach Alice Springs, and food could only be obtained at outback hotels located at the few stops along the way. Those who travelled on the Dirty Ghan soon figured out that publicans were paying the train crew to leave early – sounding the whistle as meals were being served – then pocketing the money and waving farewell.

The approach to Alice Springs saw the black, smoke-belching locomotive shuffling along the valley floor of Heavitree Gap, a spectacular red-sandstone canyon carved out by water 300 million years ago, its towering stone walls dwarfing the narrow-gauge railway snaking below. Blakeley never tired of seeing Heavitree, and this was his first time to the Alice by train – luxury compared with riding a bicycle.

This spectacular southern entranceway through the gap signalled the train's imminent termination at the railway station at Alice Springs – the end of the line from Oodnadatta.

The remainder of the expedition arrived in the Alice piecemeal, or as Blakeley put it, 'in dribs and drabs'. Blakiston-Houston preferred to make his own way there, booking a first-class ticket on the passenger train. Errol Coote in the *Golden Quest* took off from Sydney's Mascot airport on 19 July, flying over the Blue Mountains before landing in Parkes to pick up Phil Taylor, the mechanic. Their flight path would join the dots to some of the most spectacularly remote outposts in the outback – Lake Frome, Marree, Lake Eyre, Oodnadatta – until they would be in striking distance of Alice Springs. Whatever anyone thought of the bumptious Coote, his flight to the dead heart in 1930 would be an impressive feat.

•

The town of Alice Springs had all the geographical remoteness of fabled romantic destinations like Timbuktu or Xanadu, with little of the fabled romance. If anything, Alice Springs – or Stuart as the town was still technically gazetted – was as hard if not harder than any town in the American Wild West. Whichever direction you headed from the Alice, you were stepping off into the wilderness. There was no road north, only the camel track following the telegraph line. You could, if you were brave, risk following the track for 1000 miles to Darwin. The first car to accomplish the journey barely did so in 1908, and even by 1930 only a handful of people had achieved the distance.

The CAGE expedition arrived in the Alice during a unique – albeit brief – period in the outpost's history. For some time, the federal government believed the Northern Territory was too big and diverse an area for one regional authority to administer. Between 1927 and 1931, a separate territory was established, with Alice Springs as its centre. The area stretching from the South Australian border to Tennant Creek was known as the Territory of Central Australia – or colloquially as Centralia – administered by an appointed Government Resident.

In 1930 there wasn't much at Alice Springs – Wallis and Foggarty's store, Phil Windle's garage, the telegraph station, the Australian Inland Mission, the Stuart Arms Hotel, the police station, Stuart Town Gaol, a few stores, a few simple houses and that was about it. The main street, Todd Street, was simply a stretch of dirt. In the summer months, its red dust surface baked hard and dry in 45-degree heat. The Todd River that ran through Heavitree Gap and then through town was bone dry for the best part of the year, and in 1930 Central Australia hadn't seen rain for seven years.

In many ways the Alice was still more nineteenth than twentieth century. Cars in the Alice were somewhat of a rarity, if not a novelty, as petrol, imported in drums and sold in square tins, was hard to come by and expensive. Horses and camels were still the transport of choice.

Curiously, Alice Springs largely owed its existence to the very ruby rush Lasseter had spoken of. In 1883, David Lindsay, an explorer-surveyor, found what he thought were the blood-red gemstones in the McDonnell Ranges. Once word was out, miners followed the telegraph line from Adelaide through Heavitree Gap and set up camp on the Todd River. The rubies turned out to be exactly as Lasseter said: garnets. Nevertheless, miners continued to arrive and fossick through this largely untouched region. It was so promising that the South Australian government sent Lindsay back to set up a town, and in 1889 it was proposed to build a railway north from Port Augusta connecting to Alice Springs, or Stuart as it was then known. But in the early 1890s governments changed and the ruby rush collapsed, the railway not materialising for almost another 40 years.

The completion of the railway line from Oodnadatta to Alice Springs in 1929 saw the town's white population grow from 40 to around 200. At about the same time as Lasseter, Blakeley and Sutherland arrived by cattle train, so did the very first organised tourist group to the outpost town. It was difficult to appreciate just how vast Alice Springs' reach through seemingly empty regions was; author Ion Idriess noted that local priest 'Father Long's parish was bigger than Ireland'.

Despite the publicity the CAGE expedition had generated in the big cities, the townsfolk in Alice Springs appeared quite indifferent to their arrival. Since the railway's recent completion, the locals had been inundated with treasure hunting expeditions of all shapes and sizes put together by opportunist city slickers. Blakeley recalled he 'thought they were the twenty-second or twenty-third to arrive with full organisation and plant on a similar quest for treasure'. Word around town was that other expeditions were already on the move. On arrival, Blakeley was informed that a party in Western Australia had unearthed old papers and maps revealing a Lasseter-style gold find and had set out with a team of camels. Further, news came that the well-known explorer Michael Terry 'was out with his expedition on practically the same quest'. English-born adventurer Terry was a force to be reckoned with. Whip smart and tough as nails, his love of adventure combined with his colourful writing skills saw him once described as 'a buccaneer'. During World War I he'd served in an armoured car unit with the Royal Naval Air Service, and while on active duty in Russia in the lead-up to the October Revolution in 1917, he was captured by the Bolsheviks. Eventually released, he was discharged from the navy and moved to Australia, the continent's red centre proving a perfect place for his entrepreneurial skills – and recent word of an El Dorado had seen him already underway with camels and a truck. While this no doubt worried Blakeley, it seemed to only amuse Lasseter. 'Let them all come,' he smiled wryly, '. . . if they can.'

Newspapers in the big cities were running hot with the excitement of a modern railway generating a wild new gold rush out in unknown Centralia, a welcome phenomenon about to turn a broke nation on its head. As the *Adelaide Advertiser* espoused: 'We are on the eve of a great mining revival, recalling the history of past years, and, possibly, great things. Never since the dying days of the mining industry in Australia have we witnessed such activity, and never have expectations run so high. Six expeditions with full transport equipment, including aeroplanes, motor trucks, and camel trains, backed by thousands of pounds of subscribed capital, and staffed by men of experience and distinguished ability, are on the move to open up new Eldorados in Central Australia.'

These other missions were indeed as the newspapers indicated – staffed by men of experience and distinguished ability.

However, the CAGE expedition was something else. It was in fact a hastily cobbled together company of disparate go-getters – as the locals would call them – being strung along by a member of the expedition who claimed to know the precise location of the goldfield but wasn't telling anyone where it was. Where they were heading exactly, no-one really knew – west, presumably. No-one knew, except Lasseter.

Blakeley was never quite sure about Lasseter – whether this curious, strange-looking figure who had materialised from a mundane Sydney suburb would truly arise as Australia's financial saviour as everyone said. As Blakeley had been anointed as the expedition leader charged with finding the fabled reef, he certainly hoped so. Yet in the back of his mind was always the worrying possibility that Lasseter could be some kind of charlatan or confidence trickster. Or he might be delusional, or simply deranged. One way or another, the increasingly nervous Blakeley knew he would eventually be confronting the truth somewhere out in the desert.

In Alice Springs his fears would soon take shape.

12

'WE DRAW THE COLOUR LINE HERE...'

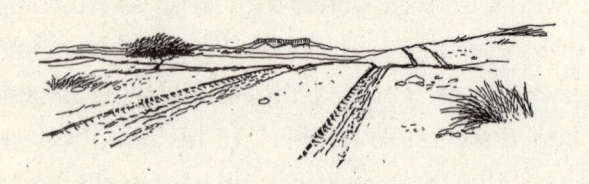

The *Golden Quest* buffeted and battled its way through countless air pockets high above the Flinders Ranges, the rugged topography below unfolding as a cruel and uninviting landscape. Errol Coote and his passenger Phil Taylor were making heavy going of it. 'There was a strong wind on our starboard quarter and the plane was thrown around violently,' Coote wrote of the journey. 'Looking down I was not reassured by the country. It seemed as barren as the ramparts that line the Red Sea, and its rugged nature as devoid of landing grounds as a chicken of its teeth.' In time they found the railway line to follow; this was not always easy, as in places it was covered with drifting sand. The journey was rough, tiring and monotonous. Eventually the few buildings that defined the town of Marree came into view. On landing, the aircraft was suddenly swamped with 'Afghans by the score', as Coote put it. A tiny Austin 7 made its way through the crowd, its driver telling him he had stupidly landed in 'Ghan Town', where whites wouldn't venture. 'We draw the colour line here,'

Coote was told. He had not seen a town divided so clearly and overtly on race. For the plane's safety he was advised to remove the aircraft and land on the racecourse.

It was not unusual to find such animosity towards the Muslim Afghans in far-flung regions. In the silver-mining town of Euriowie, 50 miles from Broken Hill, the publican so despised them that he butchered a pig, and while the Afghans were rounding up their camels, placed pieces of the animal in their camp oven, quart pots and other cooking equipment.

The following day, Coote and Taylor continued on, passing the vast salt flat of Lake Eyre and then landing at Oodnadatta to refuel. They were now on the last stage of the flight, crossing red sand dunes that stretched to the horizon, and far below the railway line came into view. Roads then appeared, signalling that Alice Springs was close. 'On the other side of the range nestled a canvas and galvanised-iron town on the banks of a dry creek with tall gum trees,' wrote Coote. 'We had arrived!'

They had flown through the spectacular entranceway that was Heavitree Gap and could now see the aerodrome. And it was then that the engine inexplicably cut out. The loud drone from the exhaust that had been punishing their ears for so long instantly vanished – silence. Coote jammed the throttle forward and the aircraft plunged into a dive, with the ground looming worryingly close. The engine then sputtered noisily back into life. Coote could see Lasseter and Blakeley running across the airfield and he eased the *Golden Quest* onto the soft runway surface. The aircraft taxied to a stop and Coote grounded the magneto, shutting down the engine. He and Taylor had just had a frighteningly close call, and why the plane suddenly stopped in mid-air he had no idea. 'A stoppage like that was alarming,' Coote worried, 'and I determined to have the trouble fixed before we went any further with the expedition.'

Nevertheless, they had arrived safely. As the pilot recalled, 'The first phase of the expedition was completed – the real job was now about to commence.'

13

STRANGE BEHAVIOUR

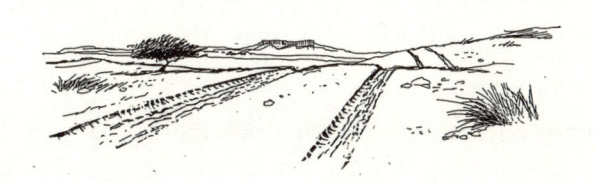

As a matter of procedure, Fred Blakeley sought to visit Alice Springs' special magistrate and postmaster Ernest Allchurch, a towering and well-respected figure in Central Australia. Blakeley had known Allchurch for more than 20 years, chiefly from his long-distance cycling days, and was fully aware it would be unwise – indeed irresponsible – to embark on a journey anywhere from Alice Springs without consulting him. In 1930, Alice Springs was an outpost to the frontier of absolutely nowhere, and anyone seeking to undertake any sort of expedition into the wilderness was required to heed the advice of local authorities. If anything went wrong, they would be the ones who had the task of rescuing you – or bringing the sun-bleached husk of your remains back in a dray. Allchurch's knowledge of the region – its topography, weather conditions, accessibility and potential dangers – was second to none among white people. To some degree Allchurch *was* Alice Springs – as postmaster and as a magistrate he knew everything that took place in and out of the town.

In the early days of the twentieth century he had been an operator on the overland telegraph to Darwin, in 1908 travelling with pioneer motorists Harry Dutton and Murray Aunger on the first successful trans-continental crossing of Australia by motor car. By an amazing coincidence, the young Fred Blakeley and the O'Neill brothers had cycled into Alice Springs as Allchurch was preparing to leave in the car with the two motorists. Dutton and Aunger had just hired Allchurch to travel with them so that nightly he could tap into the telegraph line and relay the expedition's progress 'over the wire'. This was an ingenious idea. The pair had previously attempted the drive from Adelaide to Darwin following the overland telegraph, but their English Talbot car broke down south of Tennant Creek, both motorists nearly perishing. Like Lasseter's miracle, they were rescued by an Afghan cameleer, returned to the railhead at Oodnadatta and entrained back to Adelaide. But they were more determined than ever to succeed in their objective. For their second attempt to drive across the continent the pair weren't taking any chances, inviting Allchurch the Alice Springs telegraph operator to travel with them in their latest Talbot. His ability to transmit messages along the telegraph line allowed them to maintain contact with and confirm their position to the outside world.

Over the years, Allchurch had accrued tremendous respect throughout the Alice Springs community, and as special magistrate was renowned for being particularly compassionate towards Aborigines, something unusual for the era. He had little time for the trainloads of prospecting opportunists and trouble-makers who had begun to descend on the Alice thanks to the newly opened railway. The Alice Springs special magistrate possessed a long memory and a mind like a steel trap. He was a man of few words, but every word he did speak counted.

Blakeley remembered the meeting with Allchurch that day: 'I made arrangements for him to meet Lasseter and when I took him to the telegraph station Lasseter made his first noticeable mistake.'

Allchurch was quick to realise when someone was 'trying it on'. He had seen it many times from the bench, and did not appreciate being taken for a fool.

Having been introduced to Allchurch, Lasseter suddenly became very talkative, reminiscing to the old-timer about various buildings around the town he had seen when passing through the Alice all those years ago, pointing to stables and harness shops and the general storerooms and saying he had been there 'just after the great ruby rush was over and got supplies'.

'That's a good bit over 25 years ago, mate,' replied Allchurch warily.

'Yes, I know that. I came here from Queensland to mine for rubies . . .'

Allchurch didn't mince his words. 'Well, you don't remember those buildings – they are only a little over twenty years old.'

For Blakeley, alarm bells started going off. What was Lasseter doing? 'But that was not warning enough,' Blakeley recalled. 'Lasseter went on to tell how he had come here and got fresh supplies.'

'From this station?' asked Allchurch.

'Yes,' replied Lasseter.

'Well, old man, all I can say is that you dreamt it, for I have been stationmaster here for forty years and I am sure that never once did I issue stores to anyone,' said Allchurch. He didn't hold back. 'And I think if you stayed here I would remember you, because you must have been very young.'

Lasseter's clumsy performance before Allchurch put Blakeley into a mild panic. 'I stopped the cross-examination by taking

Harry away,' Blakeley recalled, 'and got into him for talking such rot.'

Blakeley's concerns were soon compounded further when he discovered that Lasseter was unaware of even simple things about Alice Springs – things anyone who had ever been there knew, such as the telegraph office being a full two miles from the town. Lasseter also claimed that the big store that supplied the entire community was perhaps even bigger in size when he had passed through.

Further, a local recalled talking to Lasseter while he was buying flour and asking him 'how many bags he [had] loaded to "pack" on the journey to Perth. Without blinkin' an eye, he said four fifties. No-one spoke, and the chap who asked the question just looked at him, then he turned away. Every man in the store knew that thirty years ago, flour came in 200 pound bags.'

Over a drink one night, Allchurch took Blakeley aside and gave him his opinion of Lasseter. As Blakeley recalled, Allchurch 'was very scathing and told me I was a fool to believe anything the fellow said'.

Word then came back that 'Lasseter was making a fool of himself' at the Australian Inland Mission Hostel. A social gathering at the hospital was underway when Lasseter apparently climbed on the stage, and after giving an impromptu speech about the reef, presented the hostel with one guinea, promising that it would be £1000 when the reef was discovered. Having just been informed of this, Coote raced to the hall to discover the meeting had just broken up, a smiling Lasseter standing among the throng and reassuring them, 'It's there all right, and your money will be safe. That was only a deposit.'

Coote was furious. 'Caustic comments were being passed by people near me about the fellow being "dotty",' he recalled, 'and I felt savage: he was holding up the expedition to ridicule.'

Talking with the other expedition members, Coote articulated the general feeling. "'There's one thing," I said grimly. "If he has brought us out on a wild goose chase we'll give him his rations and water and make him walk back.'"

•

Lasseter's questionable behaviour wasn't the expedition's only worry. Back in Sydney, CAGE company chairman John Bailey had made a gaffe that threw an unwanted spotlight on Blakeley's expedition in Alice Springs. In an interview with the *Sydney Morning Herald* Bailey was quizzed as to whether he expected any trouble from 'the natives'; he replied that he didn't think so, as the party was 'taking no risks and [would be] armed to the teeth'.

Bailey's statement sparked an uproar among sections of the media and the public, sounding an alarm and reaffirming suspicions that the expedition was made up of rapacious opportunists who would not think twice about using a gun in order to secure their prize. Blakeley recalled how there was a 'hue and cry of how the expedition was like a travelling fort and that we were all armed to the teeth and they heard we were going to slaughter all the natives'.

The region to which they were headed had in recent times made worldwide news concerning the murder of a white dingo scalper by three Aborigines. The killing had triggered a wave of violent reprisals conducted by a vicious rogue Northern Territory police constable, who oversaw the murder of an unknown number of Aboriginal men, women and children. Readers in the big cities were appalled. Blakeley was desperate to distance himself from white-fella killing sprees in the outback.

Things then went from bad to worse. Capitalising on the fact that Fred Blakeley, the expedition leader, was the brother of the Minister for Home Affairs, the federal opposition launched an attack on the expedition, concentrating on its dubious make-up. The wily Conservative politician Archdale Parkhill, on the warpath against the foundering Scullin Government, zeroed in on what looked and smelled like straight-out nepotism. During Question Time, Parkhill attempted to flush out the expedition's murky connections with the Scullin Government, the Australian Workers Union, its reviled boss Jack Bailey, his son and the federal minister's brother.

Arthur Blakeley handled the questions clumsily, inadvertently suggesting the federal government would hand out free freight to any prospector wanting to travel to Central Australia. It didn't look good, as Hansard recorded.

> MR ARCHDALE PARKHILL: What assistance if any has been given by the Government to a company known as the Central Australian Gold Exploration Company?
>
> MR BLAKELEY: The Government has agreed to grant free transit for one Thornycroft lorry and four men from Quorn to Alice Springs.
>
> MR ARCHDALE PARKHILL: Is that assistance available to anybody?
>
> MR BLAKELEY: Yes for the same purpose and if going to Central Australia.
>
> MR ARCHDALE PARKHILL: Can the Minister inform the House of the name of the secretary and the leader of the company?
>
> MR BLAKELEY: The secretary is Mr E. H. Bailey. I do not know what the honourable member means by the 'leader of the company' the leader of the expedition is my brother Mr F. Blakeley. So that there may be no misapprehension on the part of honourable members let me say that the Central Australian Gold Exploration Company is in exactly the same position as any other company . . .

'The same position' or not, Parkhill smelled blood in the water and then repeated John Bailey's 'armed to the teeth' proclamation to an indignant house. The humiliating result of Parkhill's inquisition was that Fred Blakeley was ordered to present himself to the resident magistrate in Alice Springs with an itemised declaration of all the firearms to be carried. Blakeley would then need to telegraph his brother in Canberra to give an inventory of the expedition's armoury. 'Each man had been issued one revolver; there were two sporting rifles and one shotgun', the leader explaining that they were needed in case of encountering a 'mad brumby camel'. Blakeley had once been bailed up by a rogue bull camel, and said that 'Had I not had my revolver handy, I might have got my wings'.

The inconvenience and degrading awkwardness of fronting the resident magistrate was part of a growing series of problems. Blakeley was particularly annoyed by John Bailey's clumsy and boorish remark, but even more so the old overlanding cyclist was incensed by the attitude of 'busy bodies in the cities', stating: 'I have always been a staunch protector of our natives and added in my declaration that I would not barter or collect any native weapons or remove anything from the graves of natives.' Nevertheless, the firearms packed for the expedition represented something more. They were in reality another dubious component for a campaign devised for which modern technology would supposedly overcome nature.

14

PREPARATIONS

It was at Alice Springs that the expedition members were first able to appraise the much-heralded Thornycroft truck, and it took only a few short test drives before everyone realised just how diabolically thirsty the big machine was. It drank fuel like there was no tomorrow. Blakeley discovered that when fully laden, the truck swallowed a gallon of petrol every three miles. He sat down to figure out how much fuel would be needed to reach the vicinity of Lasseter's reef, and ran his hand through his hair, despairing at the realisation that there was no room on board for the amount of petrol required. The truck would simply head into the desert and run out of fuel, stranded somewhere in the wilderness west of Alice Springs. What Blakeley needed was an extra truck purely for carrying fuel. In effect, the Thornycroft was there to carry fuel for Coote's plane, and now they needed another vehicle to carry fuel for the Thornycroft. No-one back in Sydney had thought this through. Besides which, Coote was supposed to be the head of the transport committee.

The big truck was presenting itself as a problem. By 1930, Henry Ford's Model T had carved out an enviable reputation for outback motoring in Australia for the best part of 20 years. While motor vehicles were still relatively scarce in Alice Springs, the Model T – affectionately nicknamed the 'Squatter's Joy' in the outback – would have been the most identifiable. It was inexpensive, robust, light, reliable, simple to maintain, agile and easy to drive through the bush, making it nothing like the Thornycroft truck. The mindset was probably that the Thornycroft would be employed on the expedition somewhat like a tank. Impervious to scrub and washouts, it would presumably crash through where no car could hope to go. But even so, those with a knowledge of the outback weren't backward in voicing their opinions of the newly unloaded truck. There was no shortage of old-timers in Alice Springs sceptical of how this machine would survive in the outback. 'The criticism of the big truck was funny to hear,' Blakeley recalled. 'Almost everyone predicted that it would not get a hundred miles. They said it was too big and heavy, and too low-set, but they did not know what a six-wheeled Thornycroft could do . . .'

This was a dangerous train of thought. It is unlikely anyone on the expedition had received any genuine driver training or was familiar with a vehicle remotely like the Thornycroft. Blakeley's stoic belief the truck was unstoppable was not only misguided, but in the hands of an inexperienced driver the machine was potentially lethal. Even on made roads the Thornycroft was more than a handful – heavy to steer, painfully slow, with cumbersome, difficult-to-change gears and non-servo-assisted mechanical brakes. It possessed a vast turning circle, and simply negotiating a corner required maximum strength from the driver, who needed both hands and flexed muscles to heave the enormous steering wheel around. And with a load of several tons, including crew members, it was a glacially paced behemoth that took an age to pick up speed and an age to stop. Across the inside of the upright timber cab ran a bench seat on which sat the driver and two passengers. With the doors closed and windows wound up, the cabin became both an inescapable sound booth and an oven, the straight-cut gears from the heat-thumping gearbox below growling and whining, making conversation virtually impossible. Communication was only accomplished by top-of-the-lung shouting accompanied by hand gestures. But it was the truck's ill-founded reputation for invincibility was where the real dangers lay, as the deceptive notion that it was indestructible could very well lead the expeditioners into catastrophe. Certainly British truck manufacturers like Thornycroft and rivals such as Crossley and AEC had experienced some success struggling across deserts in the Middle East in various British army and RAF expeditions, but they always travelled in convoy with teams of specially trained drivers and qualified mechanics. Not so for the CAGE expedition. It would be a learn-on-the-job enterprise.

•

Perhaps it was Fred Blakeley's ponderous style of team management, but once in Alice Springs it was obvious that the expedition had not exactly hit the ground running. The assembly of both the CAGE company and the ensuing expedition to find the lost reef had clearly been hurried – even reckless – in the frantic desire to locate the gold as quickly as possible. But now this sense of urgency seemed to have run out of steam. Preparations for departure seemed to be carried out at more of a shuffle. Coote was well aware that things had slowed and he was becoming increasingly frustrated as each day passed. By now the expedition should be underway, depositing fuel dumps at suitable runway sites for his aircraft. However, he observed instead: 'The Thornycroft was almost fully loaded, yet there still seemed tons of stuff to go on board . . . seeing all this extra loading lying around, I therefore made no comment.' Coote might have made no comment to Blakeley, but he had plenty to say to the company back in Sydney.

Preparations in Alice Springs, or the 'work of sorting out things' as Blakeley put it, took much more time than anyone expected. In fairness, Blakeley – the bushman cyclist – had no experience of assembling such an expedition, but the board members in Sydney were now losing patience, wanting to know why the expedition wasn't underway. 'Here is where the Sydney office show their ignorance of what a huge job it was to get the six-wheeled truck packed so that it had any chance of getting through,' Blakeley wrote.

Lasseter fussed obsessively, loading and unloading equipment on and off the truck. He seemed to busy himself, oblivious to everyone watching, heaving around his personal, padlocked tin trunk – with God only knew what packed inside – picking things up, moving them, putting them down, constantly rearranging and repacking. Coote described Lasseter's unusual compulsive habit as being 'infernally busy'.

Long-time locals watching him fuss with his trunk were sceptical of the prospector's story, doubting he had ever been bush before, with one commenting, 'From here to Perth's a long way, even for a bushman. Too bloody far for a new-come-up who . . . travels with a portmanteau. Not him, nor no man like him.'

It was the Everest-experienced Blakiston-Houston who took the reins in loading the truck properly. Timber cases filled with provisions were carefully positioned on board the Thornycroft truck, the mountain of equipment now somewhat of a concern. They were certainly well stocked, but exactly how one truck could carry it all and the entire crew was a growing problem. Six hundred gallons of petrol, 100 gallons of water and two tons of provisions and equipment were being stacked and pieced together in a giant monkey-puzzle. The food store taken with them was extensive, the menu including pork and beans, bacon, corned beef, bully beef, tinned fish, sauces, jams, oatmeal, canned peas, pressed vegetables, tinned fruit, golden syrup and condensed milk. A large, lidded cast-iron pot known as a camp oven would be used to prepare food for the party, with four tin billies and two frying pans. They also packed an 18 by 10 foot canvas tent with fly, two .32-20 calibre lever-action rifles, one double-barrelled shotgun, ammunition, blankets and crockery, wireless equipment, oil and batteries.

Blakeley even packed his bicycle – as a lifeboat – in case the truck broke down. And for Blakeley, therein lay a hitherto-unaddressed problem. They had no backup. However, he soon found a solution.

Coote, who had little patience for the ground-crew component of the operation, received a shock. To his astonishment, Blakeley blithely informed him that as leader he'd made an executive decision to employ an old Alice Springs mate who owned a truck, a local bushman named Fred Colson.

Coote exploded, furious that some pal of Blakeley's was inveigling his way into the expedition to share the spoils. 'But Colson is not a member of the company, and only shareholders are to go with the expedition,' he objected.

'I don't intend that he shall go all the way; only to our first base,' Blakeley replied. 'Then we will let him go. Anyhow, I am leader of the expedition and I have to be governed by the circumstances as I find them.'

'All right,' said an irate Coote. 'As you say, you are leader. On what terms are you engaging him?'

Blakeley explained, 'We pay him three pounds a day for his services and his truck. We supply petrol and oil. He will also include his car for that.'

'I should say so,' Coote retorted. 'I presume you will wire the company to that effect, and I suppose Lasseter is satisfied?' Angry about the lack of consultation, Coote just wanted to needle Blakeley about what the key person in the whole operation had to say. He would put money on it that Lasseter wasn't consulted either. Coote was suspicious of why this fellow Colson had suddenly appeared on the scene.

Despite Coote's objections, Blakeley's decision in engaging his mate Colson did in fact make perfect sense. Blakeley was nervous, particularly about the Thornycroft suffering a mishap hundreds of miles from anywhere and the group becoming stranded, or worse. That the company had not considered a second vehicle for the expedition showed just how little thought had been given to the team's welfare during what was transpiring as an inconsiderate and haphazard planning process.

But this made no difference to Coote, who saw Colson's inclusion as something that would dilute the big payoff when it arrived. Then again, thought Coote, perhaps Blakeley's mate was there for a more nefarious reason. Maybe Blakeley and Colson were preparing to double-cross the company when they found the reef. They were probably in cahoots.

This incident, when Blakeley felt the need to assert his authority as leader of the expedition, exposed the problem of his awkward and non-consultative relationship with his team. Had there been a discussion about Colson's inclusion, Coote might at least have felt that he'd had some input in the decision rather than just being ridden over roughshod.

From the beginning, Blakeley and Coote felt a mutual dislike. Coote viewed Blakeley as ineffectual, sometimes slow-witted and an anachronism, reluctant to embrace modern technology such as the worth of aircraft and portable wireless. The expedition's official pilot surreptitiously chiselled away at Blakeley's authority, making his boss uneasy to the point of paranoia, and would only ever begrudgingly acquiesce to any changes in plan Blakeley made.

Blakeley regarded Coote as a scurrilous journalist bent on sensationalism and as a kind of temperamental, effeminate dandy. He described him as 'a man of slight build and dark complexion, who packed short bursts of violent temper. His rigout was laughable. He wore boots like girls like to ride in at Shows, and had he been compelled to walk 10 miles he would have had only the uppers left. He did not possess the right temperament for such a trip since he was jerky and jumpy and a journalist with a vivid imagination who was always on the job. If he saw a dingo he would write of the "snarling and snapping of its teeth" . . .'

Tensions between the two would soon become worse.

There were other reasons the expedition was delayed. Much to his annoyance, Blakeley began receiving demanding telegrams from the company directors in Sydney, questioning his actions and demanding full accounts as to the decisions he had made, particularly in hiring Colson. He recalled: 'Stupid wires from headquarters came to us every day asking for explanations of what was delaying us. These annoyed me very much as they all had to be replied to and the burning of the telegraph wires was a costly matter.'

Not only was answering these telegrams costly, but it was also time-consuming. Sending a telegram was something of an ordeal: pencilling a message on a form – making every word count, as the first fourteen words cost a hefty one shilling, and twopence for every word thereafter – then handing the note to the telegraph operator who would transmit the message in Morse code to Adelaide. From there it was transmitted via the telegraph line to Sydney, where it would be transcribed and hand-delivered to the CAGE office in Elizabeth Street. It would take an hour for a message to be transmitted from the Alice to the GPO in Sydney; an urgent telegram – as they all were regarding the expedition – would cost double.

Blakeley was dumbfounded by the directors' churlish attitude towards his hiring of another truck. What didn't the mob in Sydney understand? It seemed perfectly reasonable that the expedition needed a backup vehicle and another – capable – driver. Furthermore, there was something else these city people didn't appreciate: a devastating rise in the desert's temperature was just around the corner, making it incredibly risky to rely on just one truck.

'I was frightened of the time we had,' wrote Blakeley. 'We were starting out fully six weeks too late and we would have the hot weather upon us before we got a start. Motor trucks are not made to work in such heat and loose sand – no matter what their make is – for when the ground temperature reaches 175 degrees it is almost impossible to keep water in the radiator. Of course the Sydney crowd could not understand these things, not even when I wrote our lettergrams explaining why I had to have the light auxiliary truck.'

But then the picture of why Blakeley was suddenly being harassed became staggeringly clear. Blakeley learned that a member of the expedition had been secretly sending telegrams back to the directors in Sydney, criticising his leadership and claiming the expedition was being handled ineptly. Further, there had been a suggestion to the board that Blakeley and his mate Colson were actually planning to double-cross the company when they found the gold. And, as Blakeley found out, this quisling who had been informing Sydney had also been busy in Lasseter's ear.

Blakeley was apoplectic. 'It was here the first intrigue commenced,' he wrote. 'Of course I didn't know it at the time, [but] one stupid member of the company took it on himself to supply his version – he sowed the seed that I was going to work with Colson when Lasseter found the reef, and with this foul, suspicious-minded double-crossing thing sending in his stuff, it was no wonder I failed to understand the Sydney office's replies. He was not only satisfied with that but he must sow the seeds of foul suspicion into Lasseter – when I did find out, it was a wonder I did not commit murder.'

Now he had confirmation that his leadership was being undermined, Blakeley then made sure he had the support of other expedition members, and for the moment his role as leader was kept safe. 'Through them I kept my freedom but there is one thing I do hate in this world and that is a dingo,' wrote Blakeley of the traitor in his group. But his suspicions didn't just stop there, as he was convinced his nemesis had an accomplice back in Sydney who was feeding information to the top: 'a man cannot be a dingo on his own, he must have a long-eared, empty headed mate before his foul stuff can work'.

Exactly who the dingo was Blakeley refrained from saying, but if anyone was looking for a culprit it would have to be Errol Coote; and his all-hearing 'long-eared' contact in Sydney, Jack Bailey's son, Ern.

ALICE SPRINGS, 24 JULY 1930

At last the moment had arrived for the expedition to begin turning its wheels. Blakeley described it as 'the best equipped turnout that had ever been in this country'. But it wasn't a crack-of-dawn departure – it comes as no surprise that a series of niggling problems delayed the convoy until after 2 pm. Firing the engines up,

Colson's Chevrolet car peeled down Todd Street with Coote at the wheel and Blakeley next to him, Taylor following in the Thornycroft with Blakiston-Houston and Lasseter sitting alongside on the bench seat, and at the tail end came Colson in his Chev truck nicknamed 'Sunrise', loaded with petrol and oil, and with Sutherland as passenger. Unless specified otherwise, these were the agreed travelling positions for everyone on the convoy – all had to keep an eye out for one another, accounting for each person aboard their respective vehicles.

One can only imagine what was going through the expeditioners' minds as they rolled past the few uninterested locals who happened to witness the departure. That this group of extraordinary men in their well-equipped convoy were heading into the unknown – on the verge of presenting to the world the biggest gold find in history. What they were about to do was put Australia on the map as the country that broke free from the shackles of the worst financial depression the world had ever seen. The locals in Alice Springs were of course ignorant of what these men would do for them – but the expedition's success in locating Lasseter's lost reef would no doubt change their lives forever.

The convoy had only travelled several hundred yards before it ran into trouble. In crossing the dry Todd River that separates the town and the post office, Taylor lost control of the Thornycroft, and the truck veered from the hardened and compacted river crossing into soft sand – bogged up to its axles. 'Phil who had not driven this kind of vehicle before found the steering was different,' recalled Blakeley. 'So he got across the beaten wheel tracks and of course was stuck.'

Word that the mighty Thornycroft was beached in the Todd raced around town and the locals turned out in force to watch the spectacle. It was a humiliation Blakeley could have done without.

'The whole town turned out to give a hand to get the truck back on the tracks,' recalled Coote. It took a solid hour of heaving to extricate the Thornycroft from the talcum-like sand and set it back on hard ground. The chiacking Blakeley had received from Alice Springs' old-timers a few days earlier about the big truck not being able to make 100 miles must have rung in the expedition leader's ears. It couldn't make 200 yards.

Back in convoy the three vehicles pressed on, climbing through rocky, hilly country, following the barest outline of a track, Alice Springs edging further away by the mile. That evening they set up camp at the foot of the McDonnell Ranges at a waterhole known as Pantas Well. They had covered 30 miles that afternoon.

Just after dusk, one of the expedition members stopped and looked to the horizon, saying he thought he could hear the faint hum of a car engine. In time, the erratic flash of headlights revealed an approaching car, negotiating the twisting topography of rocky knolls and washouts. To everyone's surprise it turned out to be the Government Resident, Vic Carrington, and a constable of the Territory Police who had brought with them a telegram from the Sydney office advising that there was an alteration to the wireless code. The telegram had arrived in Alice Springs not long after the expedition's departure and Carrington considered it urgent enough to chase after the convoy to deliver it.

'Wishing us Godspeed, the car, turning with a scurry of sand, raced swiftly along the trail to civilisation,' Coote recalled.

As it would transpire, the wireless code would never be used.

15

ARCHIE GILES

It was the piercing cold that kept them awake. For all the talk of burning deserts and the worry of finding water, the pain of the pre-dawn cold in Central Australia in winter was excruciating. The freezing night air cut through every layer of blanket. At 3 am none of the expedition could sleep and keeping warm was proving an impossibility. No-one could sleep, that is, except for Lasseter. He had commandeered the truck cabin, had wound the windows shut and slept along its bench seat, beside him a gun and a box of ammunition. This mild suggestion of creeping paranoia caused some discussion among the other expedition members, who commented that 'it was queer. The further we went the more precautions he took and, as he was becoming touchy, none of us spoke to him about it.'

As dawn broke, the camp came to life; all wore their heavy overcoats and warmed their numbed fingers by the fire before they commenced packing up. They were in good spirits, and George Sutherland, the cook for the day, called everyone for breakfast – a mighty serving of porridge, ham and eggs, toast and tea. This mess-hall-style ritual every morning no doubt provided them with sustenance for the day, but by the time they had a fire going and cooked enough for the men and boiled washing-up water and cleaned up pans and pots and plates and mugs and then packed it all away, it was also indicative of the leisurely pace set for the journey ahead.

It was seven thirty before the expedition got going. Blakeley questionably opted to press on without refilling from the waterhole, on the understanding that there was excellent water not too far ahead. The Thornycroft was put to work through the scrub almost like an ice-breaker crashing its way through pack ice for smaller boats to follow. 'The big Thorny broke the track,' wrote Blakeley. 'Freddy came second in his truck and the car was often a bad last.' Nevertheless, the going was passable, and they arrived at a cattle station named Hamilton Downs, where Blakeley hoped to take on 100 gallons from the in-ground water tanks. But they discovered the dams almost empty, about 18 inches of 'yellow soup' lying stagnant in the bottom. They begrudgingly filled up 40 gallons of the fetid slop to use for the truck's radiators.

It was painfully obvious to everyone that Blakeley should have refilled with water back at the overnight camp at Pantas Well. For the expedition leader and a bushman of Blakeley's experience it was particularly embarrassing. The outback rule was water is water – never let it go. There was no option but to press on another 27 miles to reach Redbank Station, a cattle property run by Archie Giles, an elderly part-Aboriginal stockman whom both Blakeley and Colson had known for many years. Blakeley had in his possession a permit issued by the Government Resident requesting the hire of an Aboriginal local from Giles as a guide.

But during this comparatively short drive the convoy struck soft, powdery sand, revealing the Thornycroft's worrying propensity for becoming bogged.

This was the first time they needed to use the coconut matting, in effect 35 yard runners of 'carpet' rolled out in front of the truck's wheels to enable it to drive over the sand. For Phil Taylor, keeping the truck aligned on the 18-inch-wide strips was nerve-racking. As Blakeley recalled, 'This was Phil's first time at guiding the big lumbering truck across matting and of course it ran off a few times. Then we went shovelling it out, sometimes jacking up a wheel. If there were any bushes handy we made a bed of them under a wheel.'

It wasn't only the problem of negotiating sand that became a worry. Blakeley was becoming increasingly anxious about the problem of finding sufficient water for both drinking and for pouring into the ever-boiling truck radiators. With every mile the need for water became more desperate. He and Colson drove ahead of the convoy towards Redbank Station in the hope that Archie Giles could spare some.

On arriving at Giles's homestead, however, Blakeley and Colson found the place boarded up and the water tanks bone dry. There had not been rain here for more than three years. The staggering dry had forced Giles to give up on the shack and set up camp at a waterhole six miles away. On foot, Blakeley and Colson followed a camel track and eventually found the old man, who despite living through harsh and terrible times (the drought had cost Giles 3000 head of breeding cattle) welcomed the pair with a smile.

Colson knew Giles well. It had been 22 years since Blakeley had seen the old cattleman, and he was delighted he remembered him. It was said that Giles was the son of the explorer Ernest Giles who had wandered through the wilds of the Northern Territory and Western Australia back in the 1870s.

Ernest Giles was a particularly tough and resourceful adventurer, at one time almost perishing in what would become the Simpson Desert. On that particular expedition, the starving explorer caught a tiny wallaby and ate it alive, saving his own life. He later reminisced fondly about the warm, writhing meal: 'It weighed about two ounces and was scarcely furnished yet with fur,' he wrote. 'The instant I saw it, like an eagle I pounced upon it and ate it, living, raw, dying – fur skin bones skull and all. The delicious taste of that creature I shall never forget.' The desert did indeed do strange things to men.

•

The waterhole where Archie Giles had set up camp was hardly better than the dams the expedition had found earlier. A year of wallowing cattle had turned the water into thick green scum, the surrounding rocks plastered with excrement from the thousands of tiny birds that bathed in it. Giles boiled the billy and the three drank beef tea made from the waterhole. Blakeley showed him the permit Carrington had issued for hiring a 'native boy' as a local guide. Giles told him he was out of luck as all his boys were away mustering cattle, but he could spare a fellow named Mickey as a guide. Blakeley was hesitant. At about 45 years of age Mickey was hardly a boy, his thick white beard testified to that, but to make matters worse, he suffered badly from sandy blight – what we now call trachoma – a contagious eye disease that made his role as a tracker and an observer particularly difficult. But Blakeley had no choice. 'As this was my last chance of getting a native, I accepted him,' Blakeley recounted. As with everything else unfolding on the expedition, whatever was intended to happen never seemed to quite work out. 'Mickey knew the country for about eighty miles ahead, but had I known that he was practically blind I would not have taken him.'

Blakeley, Colson, Giles and the expedition's near-blind guide, Mickey, returned by camel to the abandoned homestead where they found Coote sitting alone on the porch. Blakiston-Houston was wandering the creek with a rifle, hoping to have a shot at some game.

From the verandah of Giles's abandoned shack, Blakeley and Coote heard the sound of the Thornycroft's engine roaring in the distance. The noise would die away before bellowing again and again. Without saying a word everyone understood that the truck was bogged once more, and raced the Chev car back along their tracks. 'There was the Thorny,' recalled Coote, 'about two miles back, with its nose in the creek while its back wheels, racing madly, were ripping away at the bank. Bark and saplings were soon brought into service and the truck reversed out of the sand.' Giles pointed to Taylor to cross further up the creek and the lumbering truck jolted across, this time with little effort, to a spot where they had decided to set up camp.

'From here on we were going into the unknown, as far as wheel traffic was concerned,' Blakeley wrote. 'Archie, perched up on his racing camel, came part of the way with us and had a grin half way across his face, for he had no trouble keeping up with us, even at the slow walk of the camel.' Giles stayed with them that night and was delighted with the evening's dinner, being given two eggs, a thick slice of ham and bread; he had been subsisting on a diet of meat and damper.

Tired and relieved that the vehicles had made it this far, the group lay around the campfire that night beneath blazing stars as you can only find in the outback night sky, swapping stories until almost midnight.

Errol Coote noted grimly, 'It would be our last social gathering for a long time.'

•

The following day the going was horrendous. Of all the challenges the expedition encountered, the problem of negotiating the vehicles through the mulga scrub was easily the most heartbreaking. An ancient type of acacia, these stunted, iron-like trees were sometimes spread out and at other times meshed in impenetrable forests, their roots and branches spearing tyres, cracking glass and gouging creases along the bodywork.

They were no longer following any semblance of track, instead they were simply bush-bashing. The vehicles became increasingly bogged, tree branches hacked off to place under the wheels to give traction. The matting strips became so clogged with sand they trebled in weight. Coote recalled the group's despair: 'the going became so heavy that it nearly broke our hearts.'

Dry creek after creek were painfully negotiated – bogging, digging, jacking, pushing, heaving, towing. Taylor had been learning by the metre how to maintain control over the elephantine Thornycroft. Driving the machine through the bush was straight-out dangerous. If one of the truck's large diameter front wheels struck a rock, it would cause the steering wheel to spin violently out of control in the driver's hands, an action known as 'kick-back'. If this happened there was no possibility of holding it – if your thumbs were gripping the inside of the wheel, the vicious spin of the steering-wheel spokes would break them. You learned to hold the giant wheel with your thumbs pointing outwards. Taylor discovered there were countless other things to worry about when piloting the truck, such as judging angles of approach when crossing creek beds, using the low gearing for descent and ascent, and applying the dubious two-wheel mechanical brakes sparingly – their ability to pull up the heavy machine would fade the more they were used.

For those who had never ventured into Central Australia before, something else lay in wait – the extraordinary presence of flies. From the moment the sun rose above the horizon until it set, every hapless individual was relentlessly harassed by swarms of black flies, the insects crawling into nostrils, finding their way into mouths, swallowed alive and buzzing; they would try to burrow into the corner of the men's eyes where their agitated victims would claw them away with their fingers.

The hunt for water was another ever-present problem and Blakeley rode the near-blind Aboriginal guide to find any kind of creek or waterhole. Mickey said he knew of a waterhole on the other side of the McDonnells called Oonah Springs. It was a long and terrible drive, the knife-edged stones they were now crossing causing puncture after puncture. If retrieving the truck from being bogged was demoralising, repairing torn tyres and inner tubes was equally as soul-destroying – if not worse.

Changing a tyre was both laborious and dangerous, particularly when dealing with the Thornycroft's large steel wheels. Placing a jack under the truck was not always easy due to the ground surface underneath. In soft conditions the jack needed to be placed on a square of steel plate or on the blade of a square-mouthed shovel to stop it sinking under the truck's weight. On rocky ground it was often difficult to position the jack satisfactorily, meaning some under-chassis mining work to get it in place. At almost 80 kilos, a single wheel was enormously heavy to manhandle; it would take two men to manoeuvre it on and off the axle.

Known as a 'split rim' type, the wheel required strenuous work with tyre levers to prise apart a steel locking ring, releasing the tyre and inner tube. The tube would then be either patched with rubber or simply replaced, and the whole componentry of wheel and tyre reassembled. The tyre would be slowly inflated by means of a hand-operated tyre pump, the process seeming to take an eternity. With the wheel bolted back on and the jack removed, the truck was ready to push on – until the next puncture, when the entire process would be revisited.

On arriving at Oonah Springs, the expeditioners then faced a 300-metre climb up the mountain range to reach the waterhole. Each man scaled the rocks carrying nine-gallon drums hooked onto a mulga branch placed across their shoulders as a yoke. 'Stumbling, sweating, swearing, and with hands and clothes torn by the jagged rocks, we were glad to call it a day,' Coote remarked.

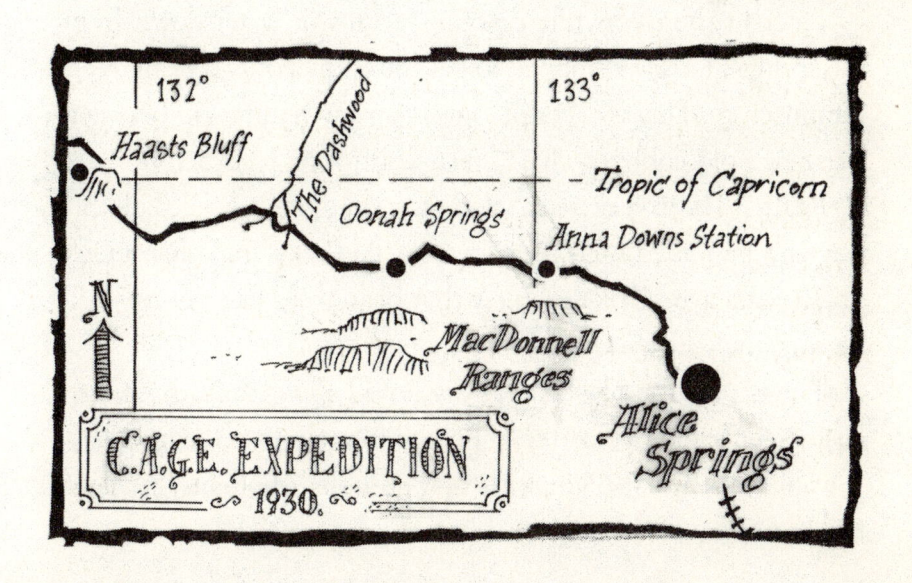

It was not lost on Blakeley that Lasseter had signed an agreement with the company back in Sydney stating that when they were 50 miles from Alice Springs he was required to notify the rest of the expedition of any landmarks he recognised. He was, after all, employed as the guide. But as the convoy drove further into the wilderness, Lasseter grew increasingly introspective and less cooperative. The drive overland had proven more difficult than anyone had imagined, and any sort of guidance from the taciturn prospector would be welcome. In fairness, Lasseter hadn't been through the area since 1897, and even then he was not in grand health, as he had explained in his story. But he had claimed that he had been navigating by landmarks, and if he now recognised any he was not forthcoming.

Then, out of the blue, Lasseter began recognising things. Strange things. Calling Blakeley's attention, Lasseter pointed to a large fault scarring a granite rock about 200 yards above them. 'See that cave up here – I can remember that I camped in there on my way out to the west.'

Blakeley was incredulous. Way up there? Not even a mountaineer could climb that. 'I looked at him in astonishment, and to see if he was doing any leg pulling,' he remembered, 'but he seemed perfectly serious and I decided to hold my tongue.'

Sutherland was also party to the discussion; he grinned knowingly at Blakeley as Lasseter walked away, 'and afterwards asked me what I thought of the statement, since we both knew that nothing but an eagle ever camped up there'.

Later in the day, while carrying drums of water down from the springs, Lasseter stopped to pat the trunks of two bean trees, reminiscing, 'This is where I rigged up my hammock, between these two old cobbers when I passed through here before . . .'

Really? Those very trees? Had the expedition travelled all this way and stumbled onto the two trees he had slung a hammock on 30 years ago? Blakeley knew that bean trees like these would be extremely lucky to last more than about 17 years.

The expedition's leader was disturbed about these 'sightings' of Lasseter's and decided to approach him about it the next day. 'I went to bed that night with very troubled thoughts,' he later recalled.

The following morning, while Taylor was working on the big truck, Blakeley called Lasseter aside and questioned him about his curious claims the day before. How was it he had no memory of the striking country they had just driven through – valleys, mountains, cliffs, overhangs, creeks – and yet he could recall some obscure crack high up in a rock and a couple of trees 'where he'd rigged his hammock' 30 years ago?

Lasseter exploded. 'You think you know everything!' he screamed at Blakeley. When Blakeley told him the trees he 'remembered' were not more than 15 years old, Lasseter exclaimed that his two trees were identical.

Blakeley had had enough and decided to put the ball squarely in Lasseter's court. '"Well," I said, "since you believe that, tell me something that is ahead today."'

They were heading into country where the McDonnells merge into another range and the landscape would change dramatically. Lasseter would not buy into Blakeley's test – it was obvious it was intended to goad him into complying with his contractual agreement as a guide to reveal whatever landmarks he recognised to the expedition.

As the convoy pushed on he sat sulking in the cab of the Thornycroft, stinging from the comments Blakeley had made about not remembering such striking country while the truck crawled through it. Lasseter hadn't finished yet.

16

THE DASHWOOD

There is an uneasy phenomenon experienced by ocean-going sailors where the moment they lose sight of land, the confines of their boat suddenly become their entire world. So it was for the CAGE expeditioners: since the moment the little convoy lost sight of the last homestead, their entire existence was held within the confines of a motor vehicle. The sheer enormity of the sprawling landscape they were attempting to conquer was now uncomfortably affecting the mood of the expedition. The open country's oppressive size was staggering. Its incomprehensible scale – the endless horizons, the distant, shimmering haze, the impossibly vast sky under which they might as well have been ants – had suddenly capsized their experience from exhilarating to harrowing. The further they travelled from civilisation, the more insignificant they had become – the stark Centralian interior had in effect swallowed them whole. They were drifting across a vast, stony, spinifex-riddled ocean, their little convoy of motor vehicles bobbing across the unbeaten tracks like dinghies in a swell.

They were headed for a large dry creek near Mount Heughlin known as Dashwood Creek. This would presumably mark the halfway point to Blakeley's first objective, the old Mackay aerodrome at Ilpbilla, and serve as a rudimentary fuel depot. The Dashwood was suggested as a waterhole by old Archie Giles back at Redbank Station, but it wasn't at all what they were hoping for. What they found on arrival was nothing more than dry yellow sand – another headache for the convoy to traverse. More chopping down of trees, more placing branches under the wheels, more digging. The heat had now noticeably increased.

Once crossed, there was the depressing realisation that there was no water here whatsoever. Alice Springs bushman Fred Colson came to the rescue, grabbing a shovel and starting to dig into the sand. Colson was making a 'soak'. Occasionally an underground bar of rock traps water when a creek dries up, the sand on top preventing it from evaporating. If you know where to look you can dig and find water in the most unlikely places.

At two feet deep the sand became moist, and a foot lower water began filling the bottom of the hole. Colson took a wooden packing case and lowered it so that buckets could be dipped in without stirring up the sand below. Errol Coote – a city slicker par excellence – was amazed at Colson's ability to produce water from nowhere. 'In ten minutes there was water as clear as crystal in the box. It tasted good too.'

Wood was cut and braced to shore up the hole as a kind of well. All the old kerosene tins and metal buckets were then filled and placed on a fire to make as much hot water as possible, as 'everyone was impatient to get his dirty clothes off', Blakeley recalled. Five of the expedition stood naked lathering themselves with soap and pouring buckets of water over each other. Mickey thought the sight hilarious, as he had never seen a naked white man before, but no amount of persuasion would make him join in.

Blakeley encouraged Taylor – filthy from conducting running repairs to the Thornycroft – to stop working with the oxy-welder and take a well-earned shower, and within 10 minutes he was enjoying buckets of hot water. Everyone was relieved to finally clean off days of sweat and grit ingrained in their skin. Everyone except Lasseter, Blakeley wrote. 'Phil was having the time of his life. But Harry would not strip off; he only washed a change of clothes.'

•

Murmurs began about Lasseter's bush skills when it came to light that he had no idea how to make the most rudimentary of bushmen's meals – damper. The recipe consisted of two ingredients – flour and water – kneaded into a slab and cooked in the flattened coals of a fire. The lump was then covered in coals for 30 minutes, only ready when the hard crust was tapped and the damper sounded hollow. It was fairly simple stuff, but when a long-distance cattleman's only rations were flour, sugar and tea, how on earth did Lasseter survive on his journey across Australia not knowing how to bake damper?

In the weeks before the expedition departed from Sydney, it was decided that at least one member of the expedition should attend wireless school to learn how to operate the portable wireless set. In 1930 a portable two-way wireless was an astonishing technological marvel, an advanced asset for the expedition. It did not require a telegraph line to tap into, which gave unprecedented flexibility when transmitting a message. Weighing a whopping 200 pounds, the 'portable' set took three struggling men to lift and carry it, who were all the while conscious the glass radio valves within were particularly fragile. The radio was powered by a supply of Eveready dry cell batteries the size and weight of house bricks, the life of each remarkably short by today's standards. In

Sydney it was Lasseter who had been nominated as the official wireless operator and technician, spending three weeks learning how to set up the machine and its aerial, its maintenance and the correct communications procedure.

Having settled in at Dashwood Creek, Blakeley decided they would make camp there for two or three days so that they could attend to running repairs on the vehicles and generally regroup before a further assault westward. This was the perfect opportunity for Blakeley to transmit his first report back to the company directors in Sydney. Together with Lasseter and the truck's mechanic Taylor, Blakeley removed the wireless from the Thornycroft, and then sent Mickey to climb a tree to hoist the aerial. When the radio was unpacked they realised there was no loudspeaker – only headphones. But that was the least of the wireless worries. Dumbfounded, they discovered the machine was missing several glass valves vital to its operation. 'This caused some caustic criticism,' wrote Coote, 'and, as the person mainly responsible for the purchase of the set, most of the comment was directed at Lasseter.'

Then, as Lasseter began connecting wires to terminals, the wireless spectacularly short-circuited, causing Taylor and Lasseter to erupt into a vicious slanging match. Blakeley quickly called Taylor away while Lasseter continued to tinker with the smoking machine. Out of the blue, he picked up the England versus Australia cricket Test broadcast from 5CL in Adelaide, which gave some hope that they would be able to transmit via a relay station at Wave Hill. Blakeley had only one call time per day – a window of 15 minutes at 9 pm – but after working on the wireless until nightfall, Lasseter and Taylor both gave up. They could receive a signal all right, but they could not transmit. The psychological impact of the radio's failure was nothing short of devastating.

The expedition's only point of contact with the outside world had been severed – their tiny domain in the dead heart had now diminished that much more. An initial feeling of disappointment transformed to helplessness. Helplessness turned to anger. As Coote remembered, 'Taylor, however, was not satisfied with the results obtained on the radio and did not hesitate to tell Lasseter so.' Taylor, the team's expert mechanic, had – like everyone else – assumed that Lasseter had checked and prepared the wireless before it was packed in Sydney. But it was all too late now. Taylor decided to take control of the radio's operation. 'This did not improve matters,' recalled Coote, 'as Taylor did not know much more about the set than Lasseter, who knew mighty little.'

Coote decided to throw in his unrequired opinion, fuelling the argument even further, until Blakeley took Lasseter aside in an attempt to calm the situation. He quietly explained to Lasseter that Coote and Taylor were 'slightly touched' in the head, and that it would be better if he just humoured them. Blakeley then wandered across to inform Coote and Taylor of what he had just said, reassuring them of his opinion towards Lasseter and explaining why he had made such a comment. 'Of course we all know that he is mad, and this is the best way out of the difficulty.'

Coote bristled at what Blakeley had suggested. 'Just as you say,' he replied, 'but I don't think the question of madness enters into the matter at all. I don't think he is mad, and I am sure he knows we are not. I think you have complicated the situation. Anyhow, we will give it a trial, but I think he is a shrewder gentleman than you give him credit for.'

•

Nights by the campfire were really the only respite anyone had. The gruelling work carried out in the ever-increasing heat was nothing short of exhausting, and so the chance to unwind by rolling a cigarette beside a bonfire under the inky night sky was particularly welcome. In those days it was common – even expected – for someone to sing or play a musical instrument or recite verse or tell a joke for what people would call a 'party trick'. Generally it was Sutherland or Taylor who would burst into song or regale the group with some hoary, lewd fireside yarn. But not Lasseter; he would not have anything to do with these brief festivities. Blakeley recalled, 'Harry had not taken part in our campfire parties. Sometimes the yarns were rough, and, as the days went on, Harry showed a distinctly religious turn.' He had, they learned, spent time in the United States, where he had become a practising Mormon.

It came as a complete surprise one night, while lying under the stars, when Lasseter began reciting tracts of Rudyard Kipling's poetry. Kipling was a popular choice for any budding impresario to recite – his stirring, imaginative works struck a chord with anglophiles all around the world. Yet while the sound of a lengthy Kipling poem could probably assist in the difficult process of falling asleep in the cold desert air, everyone lay wide awake when Lasseter unexpectedly declared that when he was a radio broadcaster, he was the individual responsible for the British author and poet's success.

'I always broadcasted Kipling,' Coote recalled Lasseter proclaiming. 'I made his works very popular.'

This was too much for Coote, who was more than cutting with his response, 'Oh, were you the cove who was responsible for that? I often wondered how Kipling became so popular.'

Lasseter failed to pick up on Coote's withering sarcasm and continued his self-appraisal.

'Yes. I specialised on Kipling over the radio, and it went down very well.' He then returned to his bed in the truck's cabin and locked the doors as he did every night.

Coote whispered to Blakiston-Houston that his caustic quip simply 'slipped out before I could stop it. I expected it to start another argument. Lasseter is a man of words, as we all are . . .'

Lasseter's outlandish boasts were beginning to wear on everyone, and showed no signs of slowing. In time, even the most resilient recipient could no longer endure his extraordinary claims. 'Oh go away! You annoy me!' Coote recalled Blakiston-Houston snapping at Lasseter during a lunch break. His patience with the mercurial expedition guide had worn through, and the Governor-General's aide had had enough. Lasseter's behaviour was indeed strange, but because he held the secret of the reef's whereabouts, the other expedition members had mostly held their tongue. However, there came a point at which his relentless self-aggrandising became too exasperating. Blakiston-Houston stormed towards Coote. 'That man's a lunatic. He claims to have designed the Harbour Bridge.'

'You think I'm a blasted liar!' shouted Lasseter. 'Here's the proof!' And he waved a page from a magazine. Doubts as to Lasseter's credibility were starting to leach through the expedition. Not only did he proclaim he had discovered the largest gold find in history, and not only had he single-handedly made the works of Rudyard Kipling popular, but he was now claiming responsibility for the greatest engineering feat Australia had ever seen.

Coote later related the Blakiston-Houston Harbour Bridge incident to Sutherland. Perceptively, Sutherland advised him to humour Lasseter, as it seemed he was angling for a reason to force the expedition to disband. 'He's cranky all right,' said Sutherland. 'Don't give him a peg to hang his hat on or he'll probably throw in the sponge.'

The days were now beginning to meld from one to another and courtesies began to fray. Blakeley and Lasseter's relationship was souring rapidly. 'The only member of the party to whine was

Lasseter,' recalled Blakeley. 'He never missed an opportunity to remind me that I was not managing the water question very well, and saying that if it did not improve, he would have to reconsider his position.'

Talk of 'managing the water question' was simply Lasseter throwing his weight around. Blakeley might have been expedition leader, but Lasseter was jerking the strings to make sure he remembered who was actually in control. It was a dangerous game. If the only person who knew the whereabouts of the greatest gold find in history suddenly stormed off, derailing the entire expedition, a furious John Bailey would want to know why. Lasseter would simply point the finger at Blakeley as the reason for the mission's failure. As the expedition leader was now aware, Coote had already stuck the shiv into his reputation back in Sydney. Blakeley was now under tremendous pressure.

It was Sutherland who would regularly defuse these hostile confrontations. 'George was a rock to me in these little passes,' recalled Blakeley. 'He would butt in and lead me away and warn me not to answer him, for he could see that Lasseter would make any old excuse to break up the party.'

17

THE AIRSTRIP

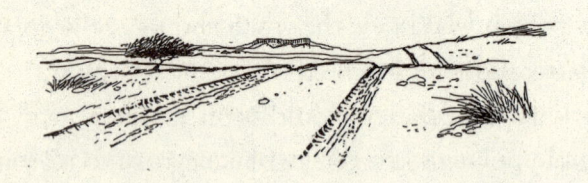

It was decided to establish a rudimentary airstrip at the Dashwood should Coote need it when he ultimately returned to Alice Springs to fly the *Golden Quest* back. Blakeley, Blakiston-Houston, Mickey the Aboriginal guide and Coote found a suitable site about half a mile away and began clearing it of saltbush. Fuel drums were placed on crude stands made of mulga wood to keep them off the ground.

Being able to give the directors back in Sydney the precise location of the newly cleared airstrip was critical – that is if they could get the wireless working. Accurate bearings were required to be taken with the sextant at noon. Yet Blakeley was sceptical of Lasseter's ability to do so, and with good reason. As everyone was aware, 30 years before, Lasseter and the professional surveyor Harding had bungled their sextant readings for the reef's location due to their watches being more than an hour slow. At the meeting in Sydney, Charles Ulm had figured out that Lasseter's positioning placed the reef somewhere in the Indian Ocean. Among the many achievements Lasseter professed during those preliminary gatherings was his qualification as a surveyor – he possessed an American Diploma of Surveying which he'd completed while he was living in the United States. That it was ordered from a correspondence school probably wasn't known at the time, but it certainly would hardly have surprised Blakeley.

Blakeley asked Captain Blakiston-Houston to accompany Lasseter, much to his irritation, to calculate their position. As it happened, Blakiston-Houston had brought with him an extraordinary navigational handbook known as *Brown's Nautical Almanac*, regarded since the 1870s as the 'sailor's bible'. It was an annually issued guide to be used with a sextant, describing the positions of particular stars and 'celestial bodies', which allowed navigators to determine their position. The almanac listed the position on the Earth's surface as to where the sun, moon, planets and bodies of stars were for each hour of the year. Looking through the sextant, a navigator measured the height of the sun, stars or planets above the horizon by aligning a combination of mirrors, then read the angle off the instrument's calibrated edge. This angle was then compared with the measurement listed in the almanac and with simple arithmetic the location could be ascertained.

The book itself was probably of limited use in this regard, as its figures were years out of date, but it gave invaluable advice as to the science of taking simple bearings with a sextant. More interestingly, this almanac had a fascinating provenance. It had been loaned to Blakiston-Houston by Rear Admiral Edward Ratcliffe Garth Russell Evans RN, KCB, DSO, SGM, 1st Baron Mountevans, who was the newly appointed commander of the Royal Australian Navy Squadron. Admiral Evans had taken this very book with him when he was second in command of Robert Falcon Scott's ill-fated expedition to the Antarctic in 1913, and its historical value could only be imagined. Evans had developed scurvy during the expedition, a condition that inadvertently saved his life. Reluctantly he did not continue to the Pole with the rest of the team – all of whom then perished. Perhaps it was this extraordinary tale that sparked Lasseter's obsession with the book: it had avoided catastrophe. While in Alice Springs, Lasseter borrowed the almanac from Blakiston-Houston and would study it at every opportunity.

After the bearing had been taken, Blakiston-Houston consulted Admiral Evans's book, only to decipher that Lasseter's position was way off the mark. Lasseter and Blakiston-Houston broke into an argument, but Blakeley's problem was of whom to believe – the correspondence school surveyor, or the captain of the Royal Hussars and Himalayas mountaineer? 'There was a good deal of heat in the argument,' wrote Blakeley, 'and when the captain quoted a passage [from the book] that indicated to his mind such a bearing could not be, Harry became sulky and would not finish the discussion.'

Lasseter's moods became more unpredictable, and minor impositions could become major issues. In repacking the Thornycroft, Blakeley asked Blakiston-Houston to remove a box of ammunition sitting on the floor of the truck's cab and place it in the back near the tailgate. Lasseter, who slept locked inside the cab every night with a loaded gun beside the ammunition box, immediately returned it, whereupon another loud argument erupted between him and Blakiston-Houston. The pair faced off as Blakeley walked towards them, telling Lasseter he wanted the ammunition in the back of the truck. The rest of the camp realised trouble was brewing and came to a halt. Coote then decided to barge in and have his say. 'How he came into the argument I cannot remember,' said Blakeley, 'so I shut him up by saying to him quietly "You are not helping, you are only making matters worse . . . you can see that Harry is half mad and for the peace of the party try to fall in with the others."' Blakeley, attempting to break up the fight, was surprised when 'Harry turned on me and accused me of hobnobbing with the aide-de-camp and asked me what would happen if we were "attacked" suddenly.' This was too much for Blakeley, who instructed Lasseter that the firearms and ammunition were for 'sporting purposes only' and, until he found the reef, 'there was not much chance of using it . . . If you follow my instructions on this trip, you will never need to pull a gun on a native.'

But Blakeley realised there was more to Lasseter's anxiety about coming under attack from Aborigines. Something wasn't right. For someone who had supposedly walked across the continent, Lasseter seemed to have little understanding or experience of Aborigines. He put the question to him.

'Why are you behaving like a new-chum that has never been out here before? One thing that is very conspicuous in all your stories is that not once did you ever tell of any contact with natives, and here you are with three noisy motors and seven other men, trying to put up an argument that we may be ambushed.'

For Blakiston-Houston the argument with Lasseter was the last straw. The Sandhurst-trained Royal Hussar and Tank Corps veteran was no fool, recognising that the fractious relations between the expedition members were deteriorating irrevocably. From now on he would have nothing more to do with the old prospector. He would give up his place aboard the Thornycroft and travel with Colson in the Chev truck.

•

Another bitingly cold pre-dawn had ensured no-one was asleep. The previous night had been particularly frustrating for Blakeley, who had decided to examine 'that bum wireless plant', as he put it, desperate to try to make contact with the outside world. He had no idea how to fix it. The radio 'had a board with over a hundred holes in it', he explained. 'Each hole was supposed to have a plug, each hole and plug were numbered, not straight-out numeral numbers, but things like 1A or KE or G to I . . .' His verdict was ultimately that 'the company was sold a pup right enough'.

The convoy set off that morning in its usual fashion, but after about five miles struck a patch of large gibber stones sitting on soft ground, causing enormous delay, as each stone had to be taken from under the coconut matting before the vehicles could roll any further. This was just another in a series of unexpected obstacles, Blakeley acknowledging the inadequacy of the Thornycroft's ground clearance – with only nine inches between the track and the differential, it was much too low for the terrain they were covering. Thick grass would gather up into a bundle beneath the truck and wrap around the spinning tail shaft before seizing it, bringing the truck to a violent stop. Sometimes the coconut matting became entangled in the whole mess, taking more than half an hour to clear.

After covering little more than 15 miles for the whole day, the group made camp. Taylor and Blakiston-Houston had taken the .32-20 calibre lever-action rifles to try their hand at hunting and had opened fire on a mob of kangaroos, Blakeley describing the event as a 'slaughter'. One kangaroo was butchered for the evening meal, with Blakeley – a former abattoir worker – carving off steaks and frying them up with onions.

It seems remarkable now that during the Great Depression, when families dined on miserable rabbits bought from the back of a dray lumbering through the alleyways of Surry Hills and Fitzroy, plentiful and far superior kangaroo meat was instead hacked up and fed to dogs. However, the expedition members – and Australians in general – weren't comfortable with the idea of eating kangaroo. Unless it was served up as impossibly expensive and exotic kangaroo-tail soup in a swank city hotel, it was pet food. But a bushman like Blakeley – or like Lasseter, who had survived alone in the wilds of Central Australia – would know of the delicious qualities of kangaroo meat. 'This is where I watched Harry closely,' recalled Blakeley, 'and he did exactly as the others who had not eaten the hopper meat. They took the tiniest bit to taste, then discovered that it was a lot better than a lot of so-called rump steak they had eaten. I felt pretty sure that this was Harry's first feed of 'roo, although he told me tales of how he had lived on it for months.'

18

HAASTS BLUFF

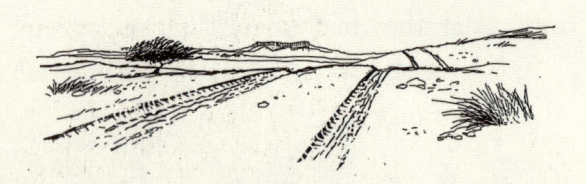

It provided some sense of achievement – and relief – when the convoy glimpsed the towering granite crag known as Haasts Bluff, a monolith situated at the start of the Belt Range, an east–west mountain range that sits squarely on the Tropic of Capricorn. An unmistakable landmark far in the distance, the Bluff served as a ready marker for keeping a straight line. Yet tight clumps of trees saw the three vehicles meandering for hours. Haasts Bluff was their objective, but to reach it they first needed to traverse the dry, deep Derwent River. Axes flailed into the mulga, and chopped branches stretched across the ground 'corduroyed' a path for the vehicles' descent from the riverbank and up the other side.

Coote recalled, 'The Chevrolet truck and the car got across without the help of the corduroy, but when the big Thornycroft lumbered down, the bank gave way beneath it, the truck falling aslant of the convoy.' There the machine lay ungracefully and helpless, with tons of supplies on board having shifted during the fall. Patience with the big truck was wearing thin; more than four hours of digging, 'sweating and swearing' saw the Thornycroft eventually extricated and pushed across the river. 'The going was murderous and we really cursed that truck,' remembered Coote.

The majestic, soaring pillar of Haasts Bluff outweighed and overcame the agony of pushing and pulling three cumbersome vehicles through the scrub. Sunset illuminated the towering outcrop, now swathed in brilliant shafts of gold and purple. 'The sun was now setting and its rays were playing on the sides of the mitre-like peak of the Bluff, which shone like burnished gold. The beauty of the scene took our breath away,' wrote Coote, whose hatred of the Thornycroft had faded for the moment. 'Gone was all thought of the truck . . . it was the most awe inspiring sight we had seen.'

The Bluff's staggering presence seemed to have an almost mesmeric effect on everyone. 'I don't think any of us will forget the sunset that evening,' remarked Blakeley. 'I am sure that even in Egypt there is not a more wonderful sight.'

However, despite all of Haasts Bluff's spectacular beauty, the core of the expedition was about to turn ugly. There must have been something peculiar about the place that evening, as the brilliance of the desert sunset was soon smothered by an oppressive darkness. Coote recalled a distinct air of uneasiness. 'That night after tea we settled down for our smoke but the stillness of the bush depressed us. It seemed to envelop us – there was not a movement; not a sound. Everything was deathly still.'

The following morning, an argument developed between Lasseter and Blakeley as to how to proceed around the monolith. Lasseter insisted they head to the left of the Bluff and south of the Belt Range, as he recalled that this was the route he had taken to reach his gold reef 30 years earlier. But Blakeley was determined to steer the convoy to the right, along the range's northern side, as this was the route specified by Ernest Allchurch, the postmaster back in Alice Springs. Blakeley reminded Lasseter that the plan had always been to make their base camp at Ilpbilla, and the northern route was the shortest possible distance.

Lasseter turned on Blakeley, announcing that Allchurch had no more idea of the reef's location than anyone else. Blakeley dug in his heels. He was, after all, the expedition leader, and with no working wireless and no way to communicate with the outside world, he wasn't going to steer the expedition off in a direction where a rescue party would never find them. Besides, there was water at Ilpbilla, and it was the place where they knew they could land the aircraft.

Whatever Lasseter thought, there was no question the expedition would follow Blakeley's route along the northern side of the

Belt Range. Nevertheless, Coote, Blakiston-Houston and Mickey the Aboriginal guide decided to drive the car along the southern side for a brief reconnaissance. The going seemed surprisingly easy, with not much more than saltbush as an impediment. In time, they swung the car around and returned to the previous night's campsite, only to discover that both trucks had gone, and so they sped off, following the tracks to catch up with the convoy.

Then, in the middle of nowhere they were confronted by the most extraordinary sight. 'We followed their tracks and were just passing a spur of the Bluff when we saw a white man waving his arms,' remembered Coote. 'We pulled in closer and to our astonishment saw that it was Blakeley.'

Blakeley, who had been standing on a rock, frantically trying to flag the car down, was clearly enraged and for the moment unable to catch his breath, clutching his chest. 'That damned fool Taylor saw me,' he gasped. 'I waved to him. He knew I had gone ahead to find a track. I waved to him . . . he waved back and drove on.'

Blakeley jumped into the car next to Coote and the four men roared off in pursuit of the two trucks – now long gone – while the expedition leader tried to explain how and why he had been cut loose in the desert.

On foot, Blakeley had wandered ahead to find a course for the two trucks to pass between sand ridges; having done so, he signalled to Colson to bring the Chev truck through first with Taylor following in the Thornycroft. But Taylor had inadvertently driven up a parallel channel, overtaking Colson's truck, and believing the other vehicles were now behind him, kept the momentum going through the sand. Colson, Sutherland and Lasseter in the Chev weren't about to lose momentum either, and gunned their way forward, as motion was everything. But in the frenzy of roaring engines and wheels spinning in the sand, the team leader was inadvertently left behind. Blakeley had furiously waved his hat at Taylor to stop but 'he only waved his hand at me and gaily kept on going as he was in top gear and had his foot hard down on the accelerator'. Breaking out onto a hard claypan, Taylor gave the Thornycroft everything he could, and for the first time during the entire journey the truck was sailing in its highest gear, barrelling across a surface 'as smooth as a billiard table and almost as level'.

The only problem was that Taylor was heading in completely the wrong direction.

Then began a wild chase across the claypan with the Thornycroft 'going like the wind', as Blakeley put it, followed by Colson's truck, chased by the Chevrolet car. After 10 miles the car overtook Colson's truck with Lasseter and Sutherland on board, and realising Taylor was the only person in the Thornycroft the chase resumed at even greater speed. Blakeley hung from the window of the car, screaming at the top of his lungs as they approached the runaway Thornycroft, while Lasseter, in the back of the Chev truck with Sutherland, produced his revolver and fired shots into the air 'to attract more attention'. Taylor eventually realised he was the focus of considerable commotion and hauled the Thornycroft to a stop. Blakeley leapt out of the car and confronted Taylor. 'Where do you think you're going?' he bellowed.

Nervously, the young mechanic enquired, 'Why, isn't this the way?'

Blakeley exploded, summoning every expletive he could think of, the other expedition members taken aback by the force of his language.

The convoy turned around and drove back to try to pick up their route.

Later, Taylor meekly approached Blakeley about the incident, still stinging from the tirade. 'My word, you spoke rough to me, as though you meant it, too.'

Blakeley's response was telling. Despite the Keystone Cops element of a car chase across the desert, a simple, thoughtless slip-up on this journey could see one or more expedition members left in the desert to die. 'Where did you think Harry and I were?' he asked. It was a good point. 'You know that our position is in your truck.'

Coote, who was revelling in Blakeley's humiliation, found the whole situation hilarious. 'That's all right,' replied Blakeley, 'but supposing you had been left in this place without food or water. You wouldn't like it.' Coote, as it happened, would unfortunately find out.

•

The mulga alongside the Bluff was the thickest they had seen. A cut-down branch would require at least two men to drag it out of the Thornycroft's path. Everyone – even the drivers – was hard at work with an axe or a mattock. They would clear one mile at a time and walk back to bring the trucks up. Worried about a sudden attack from unseen Aborigines lurking in the scrub, Lasseter voiced his concern about the long distance between the clearing gang and the vehicles. The conditions were clearly taking their toll on the two trucks and the car, and on Taylor the mechanic, who was becoming exhausted, every night making running repairs until well after dark. It was decided from then on to make camp earlier in the day: 'It was a far better arrangement,' remembered Blakeley. 'We had all the repair work done before dark and, as there was no damper to cook, we had an evening of leisure and were able to enjoy a game of cards.' The remains of the day were their only respite.

At every sunrise, the blood-red line of dawn delineating the earth from the sky brought with it new, revitalising hope. Perhaps today would be the day Lasseter would confirm the landmarks etched in his memory. Perhaps today would be the day they would get to see the fabled gold reef for themselves. Perhaps today would be the day they would become millionaires. But as the sun rose and the expedition was again underway, today was like every other, pushing three cumbersome vehicles across terrain for which they were never designed.

They reached a large, dry, gumtree-lined watercourse, Aiai Creek, its precipitous banks making it impossible for any vehicle to cross. The expeditioners wandered along the crumbling edge for several miles looking for a route across but returned to the stationary convoy without having found a way. Deciding to hand-dig a 60-metre-long cutting, they took up picks and shovels and set to work. By chopping down some of the mulga trees they laid the timber to corduroy a path down the embankment and up the other side. The plan was for the vehicles to gently nose over the edge and then, by gunning the motors, charge up onto the other bank. It was decided that the Chev truck and the car would cross the creek first, followed by the Thornycroft, as it would churn up the ground the most. Colson began the run-up in the Chev truck, tipped the bonnet over the edge and jolted to the bottom, then floored the accelerator to slither up the facing embankment. Then, with an unexpected grind and crunch, it was all over for Colson's truck.

'He almost got to the top of the bank when something went wrong,' Blakeley recalled. And indeed it had. The pinion in the truck's differential had stripped its teeth, rendering the vehicle completely immobile.

With the truck pushed out of the path of the Thornycroft, Taylor and Colson dismantled the differential and removed the offending part. They had no spare. This presented a serious problem in that the Chev truck had been serving as the auxiliary fuel carrier, and with it out of operation there was only the Thornycroft and the less than robust Chev car left. If they were to continue, they needed Colson's truck back in service. It was a more than a debilitating blow for Blakeley's expedition.

It was decided to send Colson and Coote back to Alice Springs in the car and the pair would fly the *Golden Quest* back with a new crown wheel and pinion. Whether there was a spare pinion lying somewhere in town was anybody's guess. Colson and Coote could be waiting for days – even weeks – for new parts to be sent up by train. While they were gone, the rest of the expedition would find and clear a suitable site to make a landing strip.

Blakiston-Houston's time with the expedition had almost run out. He was soon due to arrive in Brisbane in preparation for the Governor-General's return to England. He decided it was time to leave, and as Colson and Coote were driving back to Alice Springs, now was the perfect opportunity. Blakiston-Houston had proven immensely popular with the majority of the expedition, except for Lasseter. His unflinching work ethic always saw him at the forefront when it came to rolling out the heavy matting or swinging an axe.

While packing up his kit, Blakiston-Houston asked Lasseter if he would return Rear Admiral Evans's navigational almanac. Lasseter, clearly agitated, refused point blank.

'Lasseter had locked it away in his big tin trunk and no amount of argument would shift him,' recalled Blakeley. 'He just sat there and sulked.'

Exasperated and tired of pandering to the incessantly difficult prospector, Blakiston-Houston drew a deep breath, then explained to Lasseter that Rear Admiral Evans valued the book very much. 'Then he [Lasseter] clinched matters,' wrote Blakeley, 'for he stubbornly refused to go any further unless he had a book on navigation. Lasseter had been waiting for such a chance to show his authority. It was just sheer cussedness, for I will lay on oath that Lasseter never opened that book again. He did not want it, in fact he could not understand it.'

Frustrated with Lasseter's behaviour, Blakeley took Blakiston-Houston aside and told him he was prepared to have 'a showdown, whether it ended the expedition or not'.

The captain wouldn't hear of it, and so Blakeley 'gave him my word of honour that I would return the book myself to Admiral Evans's own hands'.

The Chev car was loaded up and Colson, Coote and Blakiston-Houston waved farewell through a cloud of dust as they began following the convoy's tracks back to Alice Springs.

No doubt the captain was glad to be out of there.

•

Blakeley noticed a distinct change in Lasseter's attitude as soon as Blakiston-Houston and Coote were taken out of the equation. For the life of him, Blakeley could not understand Lasseter's angst towards the captain, but now that he had gone Lasseter was working as hard as anyone at the laborious process of clearing an airstrip. Mickey, the Aboriginal guide, turned out to be an expert axe man, showing everyone how to remove obstinate mulga trees by digging a hole next to the trunk and cutting the stump below the surface. Particularly recalcitrant trees were burned out. The race was on to have a suitable area cleared for when Coote returned with the *Golden Quest*.

For the three in the Chev car now belting along the track back to Alice Springs, things weren't going well. Because they were retracing the path created by the Thornycroft crashing through the scrub, bushes and trees pushed over by the vehicles were now pointing towards the car, each hardened mulga branch a bayonet threatening to spear the radiator. The process of stopping the car to clear them out of the way was proving time-consuming. Colson decided to continue driving through the night, keen to get the most difficult part of the journey out of the way. Then, approaching Haasts Bluff, they lost the track altogether. When they eventually rediscovered their path, Blakiston-Houston and Colson set off back through the darkness to retrieve the car, the headlights switched on to mark its location. Coote lit a fire on the track to guide them back.

On arriving at the Dashwood, the over-tired Colson, who had been driving all night, pulled up and crashed out asleep. At daybreak a billy of tea was boiled up and the furious drive was back on. A stint of eight hours behind the wheel saw them arrive back at Pantas Well, Colson reckoning they could make the Alice on dusk. The car eventually pulled up at Alice Springs post office at 6 pm, and after picking up their mail they were in the town itself, quite the worse for wear. 'Houston looked anything but elegant in his moleskins,' recalled Coote, 'and his Van Dyke-style ginger beard was a complete disguise for the immaculate Captain of Hussars.'

At the pub that evening, Colson ran into the local garage proprietor and, for the first time since the expedition began, struck it lucky. The owner was confident he had a new crown wheel and pinion in stock for the Chev truck. This was exceptionally good news in that they didn't need to wait in Alice Springs for the parts to be brought up from Adelaide.

Blakiston-Houston still had a little time left before his furlough expired and he intended to try his hand mustering at a cattle station out at Love's Creek. He'd had enough of the ponderous expedition. 'Too bad I'm not in for the kill,' he remarked. 'I'd have liked to have stayed long enough to see some real gold, but

I'm afraid the going is too slow for me.' Coote asked the captain if he intended to ever return to the Himalayas. Blakiston-Houston suggested he might – there could even be gold in the mountains. 'I'd like to take George Sutherland along with me, too. He's a good old scout.'

The following morning, Coote and Colson set about removing the *Golden Quest* from the Government Resident's office yard. In the early afternoon, the overloaded aircraft took off, circling the town before heading west-northwest, to make for the expedition's camp at Aiai Creek. They were flying over the McDonnells in no time, Coote keen to be clear of the mountain range as quickly as possible. It was a decidedly bumpy ride, and for Colson, who had never flown before, it was traumatic. Aware that he was very likely to become airsick, he brought with him a pair of socks to use as makeshift sick bags, and at the first drop in an air pocket he began retching constantly.

Coote peeled off from the red earth track following the telegraph line to the north and settled in for his westerly flight across the Burt Plain, Mount Hay and the unmistakable silhouette of Haasts Bluff in the distance. These landmarks were of great assistance to Coote, as Colson – nominally the navigator – was too ill to be of any use. Following the vehicle tracks in the sand below, Coote recognised the airstrip they had cleared near the Dashwood, the fuel drums still sitting on their log stands, and before long Mount Liebig came into view, signalling that the camp was only a few miles ahead. He sighted smoke rising from where Blakeley and the others had chopped and burned the mulga away to create a large cross on which he was to land, but even the longest length of airstrip at 400 yards seemed particularly short. Coote shouted disparagingly to his passenger that the airstrip was 'big enough for a model plane'.

'The smoke was useless as a wind-direction indicator; it was blowing everywhere at once,' he wrote. 'Then straightening out, with the port wing perilously close to the mulga, we landed, pulling up within a very short space.'

Colson, spectacularly airsick, climbed out of the cockpit needing to lie down, announcing to everyone as they crowded around the plane, 'I'll wash the socks tomorrow.'

19

DEATH TRAP

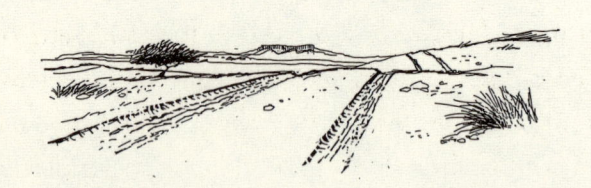

For the following morning Lasseter proposed they have breakfast in the dark, as the daylight brought with it an infuriating and relentless swarm of bush flies. These dreaded insects appeared in their millions as soon as the sun broke the skyline. Not only that, but at the same time hordes of aggressive black ants would suddenly materialise, storming across the earth in all directions. They were seemingly everywhere – taking possession of the dry, dead wood, of rocks too hot to touch, the red sand, running up the expedition members' boots and inside their trouser legs. The combination of the ants and the incessantly buzzing flies made it impossible to find any respite whatsoever. Simply sitting down or standing up was out of the question.

Nevertheless, this new day seemed promising. Overnight, Taylor, Colson and Sutherland had replaced the Chev truck's crown wheel and pinion, and the vehicle was back in operation. With the *Golden Quest* now on hand, Blakeley had asked Coote to take Lasseter on a reconnaissance flight to see if he could locate the reef. Coote, however, was nervous of the aircraft's ability to take off on such a short runway. Blakeley had told him they could have done with more time to make the strip longer, but it was too late. The pilot was faced with the unenviable task of achieving enough height to clear the tree line.

Lasseter was more than ready for the flight, rugged up in his overcoat and sporting a flying helmet and goggles. Yet Coote was still unsure about taking off – particularly with the weight of a passenger. The landing the day before had rattled him. He would have a crack at taking off, but not with Lasseter on board. He decided on a solo test flight.

'Better wait a while, until I take her out first,' he told a disappointed Lasseter. 'This ground is a bit on the small side, and there's no use two of us getting skittled . . . I'm afraid your survey flight is off.'

The expedition members manhandled the Gipsy Moth into position, pulling it well back into the scrub to give Coote as much of a run-up as possible. The runway faced directly towards the creek. Tall gum trees lining the banks towered before the idling aircraft, an intimidating barrier which Coote would either hurdle or crash into. With the engine roaring, he signalled he was ready for takeoff, with Blakeley and Taylor letting go of the wings and stepping back. Charging along the runway, Coote revved the engine beyond redline as the aircraft rose sharply. The wall of eucalypts at the creek's edge was now approaching, but the Gipsy Moth was still 10 feet too low. Coote huddled within the cockpit hoping to reduce wind resistance as he wrestled the joystick back. 'Literally, I wished her higher and higher.'

The propeller reeled skyward, Coote seeing clear wide-blue, when the plane suddenly jolted with a loud, sickening bang. In a sliver of a second, Coote braced himself for an inevitable violent impact with the ground. But it didn't happen. 'As the plane flew on I glanced to my right and saw that a small branch, snapped off a gum tree, had become hooked on the leading edge of the bottom wing and its strut.'

The ordeal was far from over, however, as the aircraft then collapsed into a nose-dive, set to spear directly into the sandy creek bed. Coote's instinctive response somehow miraculously brought the aircraft out of the fall and, spotting an opening between the trees, he flew the Gipsy Moth sideways between them, barely 10 feet from the ground. It was astonishing for everyone to see.

Coote's brush with death unnerved him, with dramatic physical results. 'Then the reaction came. The sweat began to pour from me even though it was so cold that I shivered. My legs seemingly were palsied . . .'

From the ground, Coote's spectacular aerial display had been marvellous barn-storming entertainment, but then everyone stood puzzled while the aircraft circled overhead for what seemed like hours. Coote, in fact, had almost passed out and was flying in circles to regain his nerve. It was time to face that runway again. 'Finally I felt all right and gradually moved lower and lower towards the mulga to land.'

The plane skimmed the airstrip, the wheels bouncing while the machine tore along the runway, Coote yelling at Taylor to grab the tail. The mechanic sprinted towards the aircraft but the Gipsy Moth was travelling too fast, speeding past him. It looked as if Coote was destined to plough straight into the gum trees he had scraped on takeoff. Yet he spied an opportunity, a break between the eucalypts, and 'jammed on the left rudder and held the joystick over to starboard . . . we missed the trees; but suddenly right in front of the machine was a gaping rabbit warren that had been dug out by aboriginals'. The plane somehow bounced over the void, but was now certain to fall into the riverbed. Coote cut the switches and braced himself for the collision. 'The plane's nose was pointing down the bank, when to my astonishment, she seemed to skid and then stop.' The wheels had ploughed into soft soil, bringing the aircraft to a graceful halt.

Blakeley raced up to congratulate Coote on such an astonishing feat of flying. 'Splendid!' he gushed. 'I thought you were a goner. I'll guarantee nobody could have made a more perfect landing!'

'The joke was on me,' Coote reflected on his unbelievable luck. He knew he had only just cheated death. 'If they had known I would have sold out my chance of getting away with that landing for a bad two shilling piece . . .'

The worrying issue for Coote was that the aircraft's engine had continued to rev uncontrollably despite the throttle being hauled right back. As he stood explaining the problem, fuel poured from underneath the engine, running the length of the fuselage. It was the carburettor again. 'That carburettor will be the death of me if it's not adjusted,' Coote snarled, and informed everyone that he refused to fly the plane again until Taylor had the carburettor issue sorted. It was decided that the aircraft would remain grounded there until they reached Ilpbilla, where the Mackay Aerial Expedition had cleared a proper aerodrome. In time, they would drive Coote back to fly it on. But the flight test that morning had been a frightening wake-up call as to how things could change irrevocably in a catastrophic instant. Blakeley labelled the airstrip a 'death trap'. As Coote soberly pointed out, 'Everyone agreed that had Lasseter been on board with me the machine would have been wrecked and probably both of us would have been killed.'

20

CRASH THROUGH OR CRASH

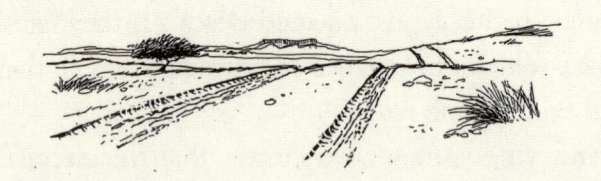

The plan now was to press on the next 100 miles to the abandoned airfield and their proposed base camp at Ilpbilla by truck, leaving the *Golden Quest* tied down, hoping neither the weather nor curious Aborigines would destroy it. Phil Taylor had patched up the fuselage holed by the gum tree, given the engine a routine check and placed a cover over the cowling. 'It looked strange to see a beautiful painted thing like the *Golden Quest* tied up to trees,' wrote Blakeley, 'and we were leaving it to take its chance with the weather.'

The mulga now had become so dense it was practically impenetrable. Blakeley and Colson tried to find a way through, climbing a tree only to discover that the entangled scrub stretched as far as the eye could see. The pair returned to bleakly inform everyone there was nothing for it but to hack their way through. With six axes swinging into the bush, the expedition inched forward, covering only three miles that morning. A discussion broke out about somehow strengthening the truck's bumper to use it like a bulldozer, driving straight at trees and pushing them over. Blakeley fashioned a hefty buffer from a myall tree, notorious for its hard timber, which Taylor then braced with other lengths at right angles running along the chassis. These he cross-braced with supports underneath. They then put the idea into action. 'It hit the first five or six trees, knocking them flying,' recalled Blakeley, and they began making headway. But the process of everyone trying to clear pushed-over trees to one side became exhausting, so Blakeley 'told Phil to get as much speed on as possible and hit them and roll them over to crush them down as the truck forged ahead'. It was a risky decision.

'Taylor was throwing caution to the wind and using the truck like an army tank,' remembered Coote. Under Blakeley's direction Taylor lined up the Thornycroft with a particularly large tree blocking its way. Blakeley recalled, 'So I said to Phil, "Drive straight at it and hit it hard." He did and knocked the tree out by the roots.' Further, it smashed the tree with such force it knocked it clear out of the truck's path. The discovery that the Thornycroft could prove an effective battering ram was particularly enlightening and so 'we started hitting everything before us'.

But the damage inflicted on the trees was reciprocated on the truck, with branches smashing the lower windscreen glass, the timber cab supports broken off and the radiator clogged solid with fine husks from a plant nicknamed 'pussycat tail', causing the Thornycroft's radiator to boil incessantly. The truck could go no further, so it was decided to set up camp. Cleaning out the radiator was particularly time-consuming for Taylor, with Mickey given the task of patching the punctured tyre inner tubes. Despite the damage, the truck's victory through the scrub seemed to reinvigorate everyone. 'We were all much better heart now, for this bumping racket and the exciting thrill of it lasted all evening,' Blakeley remembered.

The following morning, Taylor set about removing the headlights in preparation for another onslaught through the scrub. Sutherland warned everyone, 'Phil is clearing the decks for action.' Indeed he was – it was now a mission of crash through or crash. And that is precisely what happened. An onslaught aimed at a tree about a foot thick brought the Thornycroft to a violent, flying stop. 'The sudden jar threw us all up against the windscreen and gave us quite a shock. It broke both side braces and new ones had to be cut. We then discovered that both front wheels were flat, so between the repairing of braces and mending of punctures we made a new start.'

The truck was now unquestionably that much more worse for wear, and after five days of crashing through the mulga the

Thornycroft was starting to come off second best. The cab had been badly split, and flattened petrol tins were hammered to the corners to sure it up. Where the glass had smashed was now covered with timber boards. Taylor had removed the battered mudguards and was reluctantly forced to burn out the 'pussycat tail' residue blocking the radiator with a kerosene blowtorch – a procedure that could soon destroy the radiator's cooling fins. The gleaming bastion of the British motor industry was now indeed in a very sorry state.

For every inch gained battling through the bush, the mulga had been merciless and was not giving any indication it was going to relent. In fact, it was getting thicker. Small sticks as hard as iron continued to pierce the tyres, often requiring a brace-and-bit to drill them out. The steel wheels had suffered considerable damage. Severe dents now rendered them almost unserviceable and the split rims were very rough. Pulling the wheels off and splitting them apart was exhausting and dangerous.

While trying to prise apart a wheel with a four-pound hammer and tyre lever, Taylor suddenly slipped with the steel bar as Blakeley missed with the hammer, splitting the mechanic's thumb open. 'When the spanner slipped it came down on his thumb with a glancing blow. I thought the thumb was off at first. It looked a very bad smash. That ended work for the evening.'

They broke open the first-aid chest and gave Taylor a shot of brandy. 'The thumb was too ugly to investigate to see how badly it was smashed,' wrote Blakeley. 'We put the splints on and fixed up his bed.'

The following morning, Taylor informed everyone he would be able to manage, but doing any repair work was out of the question. Having the young Englishman out of action was a major setback – he was the one who kept the vehicles rolling and in one piece.

•

The slow push west gradually revealed one natural landmark after another. Before them now was Mount Liebig, a gargantuan basalt monolith towering some 1,500 yards high. Blakeley described it as 'the most spectacular mountain in Australia'. Yet for some, travelling through this region had become worrisome. Mickey the Aboriginal guide was growing increasingly agitated, confiding to Blakeley he didn't know the country from here on. There was speculation he would soon ask to leave – something Blakeley would no doubt agree to, as otherwise they would awake one morning to find him gone anyway. Mickey seemed frightened of the country they were now in. Approaching Mount Udor he refused to look at it, afraid of the 'spirits of their old gods that live in them'.

To fuel his concerns, there was a growing feeling that the convoy was being watched. Blakeley commented, 'A most remarkable sight appeared just after we started. From every direction we could see smoke signals going up. Before we started there was not a sign of one, but as soon as the old Thorny got into its low gears, up they went.'

Stopping for a lunch break at Mount Udor, Blakeley and Coote went exploring on foot. They were looking for a worn track made by Bob Buck's caravan of 70 camels, which had brought in fuel for the Mackay Aerial Expedition only a few months earlier. If they found the camel pad, that would lead them to Ilpbilla, and to water. The area at the base of Mount Udor was strewn with a jumble of colossal rocks that had sheered away from massive, brittle overhangs far above – the general feeling between Blakeley and Coote was that it was perilous to be there. Coote described the mountain as an eerie place: 'a giant spillway of boulders that would dwarf the stones of the pyramids. They have been falling from the top of the mount for centuries and clutter about its south-eastern wall to a great height.' He followed Blakeley to the top of the spillway. 'We hardly dared to breathe let alone speak, fearing that another boulder might come hurtling down upon us.' Coote was genuinely scared of the place; the sheer size of the smashed rocks frightened him and he started to panic. 'I began to run as the thought of another boulder falling struck terror into me.' And then it happened – a deafening crash as a massive shard of rock exploded on the ground only metres behind him followed by 'a sound like the earth cracking'.

When he caught up with Blakeley, Coote was white with fright, hardly able to speak, sputtering that they were 'nearly pulverised into part of the landscape'. Blakeley also felt the impact, saying he felt 'as tiny as a gnat standing under that overhang of the summit. I thought it was going to break off any minute.'

Having reached their vantage point, the pair stared silently at the panorama to the west. On the horizon, three hills announced the Ehrenberg Range. Before them, row after row of red sandhills stretched out until they melded in the distance. The idea of pushing the Thornycroft through them seemed impossible.

'The outlook seemed hopeless,' remembered Coote. 'Dirty red and pale green – the mingling of sand and spinifex – was all we could see.' But there was no option. The expedition had to keep slogging its way westward. When they set off back to the trucks, the pair realised they had lost their bearings. Blakeley felt tired and sat down, despondent that he had no clue as to how they would find Ilpbilla. Coote fired a shot in the air with his pistol and the vehicles soon appeared. Colson had been for a drive and located the elusive camel pad. Blakeley was mightily relieved. If this was correct, they were now on their way to the site. 'Even an old camel pad seemed Heaven after the last weeks of gruelling travelling through the dense scrub.'

But the hard surface of the camel pad eventually changed into fine powdery sand; when placed under the Thornycroft's wheels, the coconut matting was shot out backwards like a conveyor belt, the truck not having moved an inch. Blakeley wrote that it was 'hard to believe at first, but when it happened the second time things looked serious'. The key problem was that the truck's front wheels weren't driven. That is, the truck was not six-wheel-drive; only the rear four wheels were powered. The front wheels could not be held in a straight position no matter what, the soft sand pushing them immediately to full-lock, acting somewhat like anchors, while the rear wheels spun on the spot, spitting the mats from underneath. In time, the men figured that out by digging a trench in front of the wheels and lining it with bushes, the tyres had enough surface on which to grip and could readily climb a dune. This idea worked, but progress was staggeringly slow, making 800 yards in two and a half hours. Blakeley tumbled to another thought. 'The driving wheels were not digging in but the front wheels were,' he observed. So he dug 'spoon drains' in front of the truck to act like channels in keeping the front wheels straight while the rear wheels did all the work. It was a success. 'I was satisfied as these new tricks of nature took some thinking over.' Moving forward was a relentless exercise in problem solving.

Coote wrote of the dangers of disturbing spinifex snakes in the grass; these were 'bright green in colour, about eighteen inches long, and they coil themselves in the spinifex, from where they will strike at anything passing'. He described them as 'one of the most venomous snakes in Central Australia, and we killed dozens of them'.

In all the misery of traipsing across snake-infested desert dunes, Colson suddenly struck good luck. While following the camel pad on foot, he spied something fluttering across the ground. To his astonishment, it was a perfectly good one-pound note – an escapee from the pocket of one of the cameleers bringing fuel for the Mackay Aerial Survey Expedition at Ilpbilla a few months earlier. Aside from the delight in finding money hundreds of miles from civilisation, the note announced that they were on the right track.

•

Mickey's behaviour had become even more of a concern. He was in country where he knew he didn't belong and was frightened enough to ask Blakeley if he could camp beside the truck that night. Blakeley agreed but, because of the proximity to all things flammable, wouldn't allow him to light a fire as he normally would. Mickey didn't care. He just wanted to sleep under the truck's tarpaulin. There was danger lying in wait somewhere out there. Blakeley had reportedly seen human footprints around the camp, and was then confronted by a paranoid Lasseter and Taylor to ask if it was true. He told them it was nothing to worry about, and that from then on they would likely be seeing footprints all the time. The ritual of Lasseter locking himself in the truck's cab at night with a rifle and all the ammunition was now becoming problematic. What if there really was an emergency? What if they were attacked overnight and needed to get inside the cab? Blakeley checked on Lasseter that night, and he had indeed locked himself inside with a weapon. He goaded the old prospector about it, who responded, 'That's all right, you do as you like, but I'm not going to be caught napping.'

•

Relations between everyone were clearly breaking down, the expedition losing cohesion by the day. Blakeley, Lasseter and Coote's relationship was dissolving into a three-way impasse;

Colson, Sutherland and Taylor could only look on. Perhaps madness was taking hold.

The 100-mile journey to try to reach the Ehrenberg Ranges and the base camp at Ilpbilla had in effect become a seemingly endless nightmare – the mindless torpor of chopping a path inch by inch through impossibly dense mulga, and rescuing the vehicles from permanent graves in treacherous sand over and over again. Coote recalled, 'the sun became hotter and hotter, and the country was devoid of shade. Moreover the Ehrenbergs . . . appeared to be shifting away from us.'

They were travelling across vast stony flats, which were almost as bad as the sandhills. Coote remarked, 'Our speed was so slow that even the flies could keep pace with us, and they attacked in millions.' By the time they rested for lunch they had travelled only three miles. No-one spoke – they ate in complete silence. During this break, Taylor removed the Thornycroft's piano-hinged bonnet to try to cool down the engine and slow down the big English truck's incessant boiling.

Towards dusk, the tiny convoy was still slogging its way west, now close to the foot of the Ehrenbergs. An exhausted Taylor was slumped over the Thornycroft's steering wheel in a stupor, the truck lazily meandering around obstacles in its path. Coote, driving Colson's Chev truck in the Thornycroft's wake, suddenly sat bolt upright. Smoke belched from underneath the Thornycroft in front of him, followed by the lick of flames. Coote overtook the big truck, with Colson in the Chev's passenger seat screaming at Taylor to stop. Taylor 'jumped as if he were shot', recalled Coote, and the mechanic pulled up, wrenched a chemical fire extinguisher from the side of the truck and emptied its contents on the fire. The others raced up with shovels to heap dirt on the blaze. It was out within a few minutes. The truck's fuel tank was situated inside the cab, a rectangular metal container running the full width of the dashboard. If that had ignited, the timber cab would have become a fireball, incinerating Taylor within seconds.

'Perspiring freely, each of us looked scared,' Coote said. Scared was an understatement. If the Thornycroft – laden with petrol and mining explosives – had blown up, and if in the ensuing maelstrom the Chev truck, carrying 500 gallons of fuel, had gone up with it, then as far as everyone on the expedition was concerned that would have been it. They would be finished.

The Thornycroft's incendiary experience had proven harrowing for everybody. Coote remarked on the fire 'nearly leaving us stranded in the desert with no food. This thought together with the possibility of having no means of returning to civilisation, was the last straw to a gruelling day . . .'

The thought of what could have happened brought with it a cloud of considerable doubt as to the mission's success, and their chances of survival. The Thornycroft had become their world, and its demise would have spelled the end of everything. For those who were beginning to question Lasseter's story, the blunt realisation that they could perish chasing some whimsical fantasy came crashing home as a distinct possibility. Was it worth dying out here for what could be a wild hunt for gold? But then again, what if Lasseter's tale of a gold reef was actually true? This perennial question had the potential to send anyone mad.

•

The fire on board the truck raised tensions to another level. Taylor's decision to remove the bonnet had allowed the engine bay to be inundated with twigs, seeds and dry, dead grass – highly combustible when packed around the red-hot exhaust manifold. This, and the fact that they had still not reached the old Mackay camp at Ilpbilla, put Blakeley into a particularly bad temper. Annoyingly, Lasseter, the man responsible for everyone being out here, seemed oblivious to the gravity of what had happened that afternoon. His nightly habit of fussing with his big tin trunk packed in the back of the truck was starting to grate on everyone, particularly Blakeley.

Lasseter called out, wanting to know where his mug was, his accusatory tone suggesting someone had stolen it. Fed up, Blakeley snapped back, 'Who do you think took your mug?'

Lasseter said he wasn't saying anyone took it, whereupon Blakeley replied, 'Well, why make insinuations?' That was enough for Lasseter, who exploded, telling Blakeley he was not going to tolerate being spoken to like that. Blakeley, deciding to cool things down, told Lasseter he had been kidding.

'I know you have,' shouted Lasseter. 'And I'm not that fast asleep either. You couldn't deceive a lunatic, let alone a man with ordinary intelligence.' He hadn't finished with Blakeley, making sure he remembered who in fact was in charge. 'I have a big responsibility to the company to find this reef but I don't have to stand for you. I know that if I don't find it I'll be made to walk back. Well, I'm going to start right now. I won't go another yard with you. But before I go I'll have a piece of you.' Lasseter shaped up with his fists. 'Come on. Put them up, I'll fight you.'

For Coote, an alarm bell had gone off. The suggestion that Lasseter would be made to walk back if he didn't find the gold had been made by Coote himself. 'It was I who had made that threat before leaving Alice Springs. Someone had repeated it to Lasseter.'

Coote decided to step in, telling Lasseter and Blakeley to 'lay off'.

'What's this talk of going back?' he asked Lasseter. 'You can't do that. There's only one-way traffic on this road.'

'What do you mean?' asked Lasseter.

'Work it out for yourself,' said Coote.

Coote's entry into the argument only fuelled it further. 'The incident should have ended there,' recounted Blakeley, 'but Errol promptly chipped in, and, in ten seconds there were hot words. These two were always just on the edge of a blow-up, neither of them temperamentally fit for a trip like this. Coote completely lost his head . . . and there was some hot talk of guns.'

The potential to suddenly produce a firearm in a heated moment was always a terrifying reality. Everyone was conscious of the easy access anyone had to the expedition's weapons. Lasseter slept with a loaded gun every night. It would only take someone producing a revolver in the heat of the moment for the campsite to become a murder scene. No-one wanted to be brought back to hang.

21

ILPBILLA

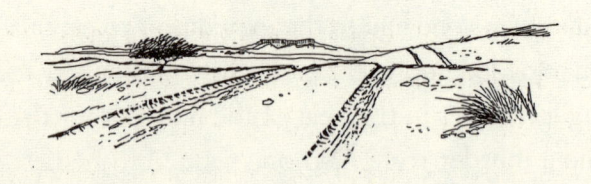

Things were still tense at breakfast. Lasseter, as Blakeley said, 'had a liver that morning and kept nagging me. He was sore about the water being scarce and accused me of not being able to handle it.' Sutherland gently advised Blakeley to steer clear of him. 'Look out, say nothing, it's only liver.'

Yet it was true about the problematic shortage of water – they had begun using their last 44-gallon drum.

Blakeley was determined to find the elusive abandoned Mackay Aerial Survey camp at Ilpbilla and the nearby waterhole that day. For everyone, locating the airstrip was vital because this outpost was a touch point with the civilised world they had left behind. Blakeley had observed the presence of a large number of topknot pigeons, signifying that water was somewhere nearby. They set off on foot, and within 20 minutes had rounded a spur to walk squarely onto the airfield. They had camped only a mile away the night before. For the expedition members the debris littering the abandoned airstrip might as well have been gold itself – discarded fuel drums, empty crates; all useless detritus, yet a welcome reminder of civilisation. Coote even discovered the frame of a Douglas motorcycle, the engine from which had been used to generate power for the Mackay expedition's wireless.

But their greatest find was shelter. Coote described the structures as 'gazebos' – two crude sheds made from branches and bushes with a separate sleeping compartment and a room containing a table and chairs fashioned from flattened petrol tins. Blakeley stood at the edge of the aerodrome and could only compare it to the amateurish effort they had hacked out at Aiai Creek some 100 miles back.

Following a camel path, Blakeley and Colson soon discovered the water supply, a limestone hole containing around 600 gallons of questionable water. There were signs of animal and bird faeces spattered around the edge, and to be on the safe side it was decided to use this water only for washing for the moment. They would continue using the drinking water brought from the Dashwood for as long as it held out. Colson drove the Chev truck up to the waterhole and he and Blakeley filled four drums while the others began to unpack the Thornycroft. Now stripped of its load, it became staggeringly clear just how badly damaged the big machine had become.

But much worse, it sounded as though the big-end bearings on the crankshaft were wearing out – if they collapsed, the engine would be finished. Taylor, his smashed thumb still in splints, knew that everyone was relying on his mechanical abilities to survive. Blakeley removed the bandage and the splints binding Taylor's thumb and the young mechanic braced himself for the filthy, painful job ahead. Blakeley offered him the help of anyone in the group but Taylor replied that it was a one-man job. Colson worked alongside on his Chev truck. Soon he would have to

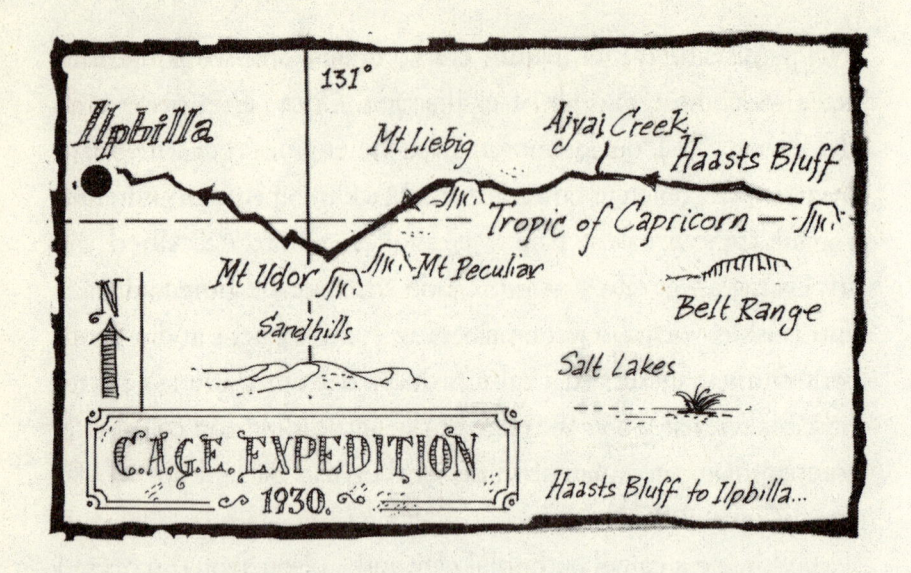

Map showing: Ilpbilla, 131°, Mt Liebig, Aiyai Creek, Haasts Bluff, Tropic of Capricorn, Mt Udor, Mt Peculiar, Sandhills, Belt Range, Salt Lakes. C.A.G.E. EXPEDITION 1930. Haasts Bluff to Ilpbilla.

undertake the long drive to Alice Springs, and then face the long drive back once more to Ilpbilla.

A fire was lit to boil water for washing. It had been two weeks since they had last bathed, their bodies filthy with grime from the desert and from the vehicles. With everyone now clean and with a few weeks' growth shaved off, it was almost as if they were starting afresh. There was plenty to do. Blakeley instructed Coote to type out a report that would be sent back to Alice Springs with Colson and posted to the directors in Sydney. Meanwhile, the inventory of supplies taken from the back of the Thornycroft was being carefully checked. When unloaded from the truck there were over three tons of supplies sitting on the ground. This was then organised into two tons of food and provisions placed in the timber shelter, and a fuel dump holding 200 gallons of petrol. Things were looking promising.

It was time to retrieve the *Golden Quest* from where they had left it tied down at Aiai Creek. Colson and Coote would drive the Chev truck 100 miles back to where the plane was anchored.

The pair would try to improve the runway. Coote would then fly the aircraft back to Ilpbilla to re-join the rest of the group and Colson would continue on to Alice Springs, send the report to Sydney, pick up the mail, fuel and water, then turn around and come back.

There were grave concerns about this plan. Coote had been extremely lucky to survive a takeoff and two landings at the 'death trap' of an airstrip at Aiai Creek. Taylor described Coote's earlier perilous launch over the treetops as 'the wildest sort of gamble, and a thing that comes off only once in a dozen times'.

Concerned for Coote's safety, Blakeley made an arrangement with him that they would wait three full days for his return. If the plane had not arrived by then, Blakeley would assume something had gone wrong and they would send out a search party. 'Take no risks,' Blakeley warned Coote. 'If you are forced down on your flight here, camp on the nearest rise and keep smoke fires going during the day and bonfires at night. As soon as three days are up we will leave here and follow our tracks back. So stick to the truck tracks as closely as you can.' That was, if he survived clearing those gum trees he had already clipped during takeoff. About the only reassuring aspect of Blakeley's idea was that Colson would be with Coote until he was in the air. The plan hardly seemed satisfactory, but it was all they had.

Blakeley gave Colson his instructions to post the report in Alice Springs, and to pick up extra fuel and 80 gallons of drinking water from the Dashwood. And with that, Colson and Coote climbed aboard the old Chev truck and swung it around to follow their tyre tracks back to Aiai Creek. Blakeley watched the truck disappear.

He felt sick to his stomach. He wondered if he had just sent Coote to his death.

•

It beggared belief that the now ragtag Central Australian Gold Exploration expedition had survived travelling such a distance, through such a brutal, unforgiving and trackless wilderness, to successfully set up an effective base camp at this far-flung outpost called Ilpbilla. It was an incredible achievement, but it had been much more arduous than anyone had imagined. It was one thing for the Mackay Aerial Survey Expedition to land an aircraft here, but it was a whole different enterprise to drive overland. Blakeley was now in preparation for the final and most crucial phase of the operation: a campaign he termed 'The Big Push', which would once and for all see the expedition get its hands on Lasseter's gold reef. With a title reminiscent of a Western Front offensive, Blakeley's 'Big Push' would be driven by speed, as the truly hot weather would begin in less than a month. From Ilpbilla they would strike out west, Lasseter guiding the expedition to his fabled reef. They would peg it, and return to Sydney victorious. Millionaires, every one of them.

But frustratingly, Lasseter's guide to where the reef lay was less than forthcoming. Yes, they would be heading off west for the 'Big Push' – but heading off to where? It was the blind leading the blind. 'So far Harry had not complied with his contract to divulge this information to me,' recalled Blakeley. 'All I managed to get out of him . . . when I took him aside and asked for the directions. He said. "Keep, first of all, a point or two north of west," showing me the position on the map of Lake Macdonald about 130 miles west of us.' Lasseter meanwhile simply busied himself trying to get the wireless to work. All anyone could do for the moment was wait for Coote to return with the *Golden Quest*, Colson to return with more fuel and Taylor to finish overhauling the Thornycroft's engine.

This time spent idly waiting was slowly having a cancerous effect on Blakeley's confidence in his own decision-making. The more they waited, the more the expedition leader kept revisiting and rethinking the hasty emergency plan he had worked out with Coote. 'I realised now the arrangement to wait three days was a bad mistake,' he recalled. If the plane crashed anywhere on the route between Aiai Creek and Ilpbilla, the pilot would be as good as dead. Even if it were forced down, without water he would die of thirst within two days, even in winter. Blakeley's doubts were gradually getting the better of him. Guilt had well and truly set in, and he was now convinced something had gone wrong. 'I had misgivings that things were not well and the three days I put in waiting were the most trying experience I ever had.' The longer they waited, the worse Blakeley's fears became. 'I could not leave the camp – reading was impossible – I was too uneasy.'

He waited for the drone of the *Golden Quest*'s engine, but it never came. At night Blakeley would sit up for hours, and at the pre-arranged times of 7.30, 8 and 9 pm he would fire the flare gun, hoping Coote could see the explosives brilliantly illuminating the black desert sky. Some of the flares burned a full 20 minutes. Taylor shone the spotlight fixed to the Thornycroft into the darkness, aiming it skyward like a searchlight. Coote's failure to appear was shaping up as the expedition's first genuine tragedy, and for the moment the gold reef had been well and truly forgotten.

Out of the blue, Lasseter announced that he had finally got the wireless to work. Surprised, Blakeley asked Lasseter to show him the machine in operation. But unbeknown to Blakeley, for hours Lasseter had been sitting with his headphones on tapping out the expedition's call sign followed by the SOS signal. When he found out, Blakeley was livid – the procedure was that only he, as the expedition leader, could authorise what was broadcast. They had made a three-day pact with Coote and still had a day to go. Sending out an SOS should be the very last resort. What if a search and rescue mission was now being mobilised due to Lasseter's tinkering with the radio? What if Coote was merely sitting by the *Golden Quest* waiting for the wind to change?

Fortunately for Blakeley, Taylor diagnosed the wireless as being dead anyway – for all Lasseter's tapping away, nothing had been broadcast. But there was more to Blakeley's anger. They were participating in a shareholder-funded exercise. If wild messages were being broadcast without his permission, this could affect the company's share price, and Blakeley knew who would end up as the scapegoat. The following morning he instructed Lasseter to pack the radio away. When he refused, Blakeley picked up an axe, walked over to Lasseter and told him, 'I'm putting this into it, then I'm sure you won't be monkeying with it while I'm away.' The wireless was packed back into its box and never used again.

By the last day, Blakeley was champing at the bit to get cracking with the search for Coote. By nightfall, the reconditioned Thornycroft had been stripped of all unnecessary gear and was ready to go. Lasseter and Mickey stayed at the camp, while Blakeley, Sutherland and Taylor set off for the 100-mile drive through the night. Blakeley described it later as 'the wildest and roughest trip ever made by motor vehicle'. With no load on the back the mighty truck bounced uncontrollably, the rear wheels 'tramping' over corrugations causing unbelievable shuddering. Blakeley shone the hand-held spotlight ahead as a third headlight, while Taylor flung the big truck through tight turns, weaving around trees and bushes. It was dangerous stuff. Sutherland was almost impaled when a thick dead branch speared through the passenger window, piercing the timber boards at the back of the cab. There had been hardly any glass left in the cabin and whatever was left was shattered, sending shards all over Blakeley, pieces of broken glass finding their way inside his shirt.

Late that night, the truck hit a tree at about 20 miles an hour, bringing it to a spectacular stop. Blakeley recalled, 'I can tell you the shock was terrible.' The hit smashed the headlights to pieces, and the three expeditioners were showered in glass flecks from the busted lenses. They climbed out, checking themselves for injuries, but fortunately all three had suffered only fairly minor cuts. The corner of the cab had been badly smashed, and at daybreak they stopped again and shored it up with a tree they cut down.

It was two in the afternoon by the time they arrived at the makeshift airstrip, and when they discovered wheel tracks where the plane had taken off Blakeley feared the worst. Taylor then spotted the aircraft about 200 metres away. Things weren't good. The beautiful black and red *Golden Quest* had crashed in the scrub off the runway, perched on its nose, upside down. The fuselage had creased at right angles at the cockpit, the wooden propeller broken off at the hub. The undercarriage, still intact, pointed upward. There was no sign of the pilot.

They pieced together what must have happened from the trail of evidence they found. Taylor pointed out dried blood in the cockpit, and from there Colson's footprints showed where he had carried the injured pilot to his truck. Everyone at least knew Coote was in safe hands with Colson taking him back to the hospital in Alice Springs. Blakeley then found attached to a stack of water drums a note Colson had written disclosing that Coote wasn't too badly hurt but had sustained a serious cut to his thigh. Colson's message indicated he would soon return, and so it was decided to dismantle the plane then and there in readiness to load it on his truck to take back to Alice Springs. As far as Blakeley was concerned, his scepticism about the aircraft's value to the expedition was vindicated. 'I never had any faith in it. I had been stuffed with all sorts of tales of its ability. It cost a thousand pounds, the mileage flown had cost the company £3.10s per mile and delayed us about 10 days, which cost another £150.' All they could do was wait for Colson to return.

22

'APPROACHING MY COUNTRY'

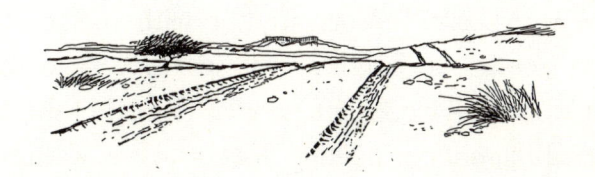

There was something indefinably unpleasant about the camp at Aiai Creek, as if the place was some malevolent vortex, perpetually dragging everyone back to face some new heartache. It was a place that had brought almost nothing but bad luck, what with stripping the Chev truck's pinion, the *Golden Quest*'s crash and Coote's brush with death. Someone suggested the that place should renamed be Disaster Creek.

That night it was almost impossible for anyone to sleep. At various times each of the crew thought they could hear the whine of Colson's Chev truck in the distance. At six o'clock they were proved correct, as Colson's battered vehicle lumbered into camp, its exhausted driver hauling himself out. Over breakfast he told them what had happened during the aircraft's ill-fated departure. On taking off, the *Golden Quest* had almost cleared the trees when the engine noise seemed to change. There was a sound 'like a pistol shot' as the right wing smashed into the branch of a dead tree. 'Then it turned over,' said Colson. 'I ran as hard as I could

157

and got him clear.' Coote's head had collided with the instrument panel, glass from the tachometer making a deep cut under his left eye. He hung upside down, blood flowing into his eyes and flying helmet. His right arm was pinned hard by the starboard wing and he had no feeling in his left leg. Suddenly, there was the calamitous realisation that petrol was dripping profusely onto the hot exhaust manifold. Conscious, a terrified Coote realised he would burn while trapped upside down. He could hear the petrol begin to sizzle. Colson raced up with an axe and swung it into the fuselage to free him from the wreckage. Within 20 minutes Colson had Coote lying on the flatbed of the truck and they headed flat-out on the long drive to Alice Springs. Coote's condition was clearly serious, with the cut to his thigh bleeding heavily. Every time they stopped, Colson noticed the blood patch getting bigger through the bandage. But 22 hours later, after a gruelling drive, Coote was delivered safely to hospital.

Everyone sat riveted, listening to Colson retell what had taken place. He had brought back with him a sizeable quantity of mail from the company in Sydney issuing numerous instructions. It seemed from these lettergrams that there was a mild panic that the expedition had sparked a new gold rush in the west of the Northern Territory and there were concerns that 'half Australia was rushing to "peg out" claims before we got there'. The four men looked at each other before scanning the horizon around their miserable locale. Who on earth would rush out here? Included in the pile of mail was a letter from Coote addressed to Blakeley urging him to contact Sydney to bring up another plane. Why doesn't he just lie there in hospital and keep out of things? Blakeley thought. Look at how much time and money had been wasted with the infernal *Golden Quest*. Blakeley hated that aircraft. He believed the only way to find the reef was to travel overland, by truck, or by bicycle or camel or foot if you had to.

He had already written a report regarding the aircraft crash for Colson to send back, and in a postscript he mentioned Coote's request. But as far as Blakeley was concerned what they needed to find the reef was not aircraft, but stout walking boots and fast-riding camels. The old-timers back in Alice Springs were right.

Blakeley, committed to having a second attempt at his 'Big Push' once he got back to Ilpbilla, made plans with Colson as to where they should meet when he returned. That afternoon, the four men loaded the crumpled remains of the *Golden Quest* onto the back of the Chev truck and roped it down, about 10 feet of the fuselage hanging over the end. At dawn, Colson, the Chev truck and the wreck of the *Golden Quest* were on their way back to the Alice.

•

Because the newly crushed track between Aiai Creek and Ilpbilla had now been traversed backwards and forwards several times, the return journey to the old Mackay Aerial Survey camp was remarkably trouble free. On arrival, Blakeley informed Lasseter of Coote and Colson's harrowing experience. Lasseter agreed with Blakeley about his 'no more planes' request to Sydney, but suggested that an aircraft could work out if only they had a different pilot. He hoped Coote wouldn't return.

With now well over a week lost, Blakeley was determined to reinstate his plans for his 'Big Push'. In preparation for the journey west, Taylor and Sutherland began servicing and refitting the mangled Thornycroft truck as best they could, knowing what the consequences would be if something went wrong. If the big truck broke down out here, there was no backup. They removed the splintered and broken boards that formed the cab, replacing them with timber from knocked-down packing cases. Taylor removed the truck's enormous mudguards, making it easier for him to see the front wheels when guiding them over the matting. They built a timber compartment to hold water and petrol drums on the back of the truck, and their supplies were labelled and packed on board.

In the early afternoon the Thornycroft was idling and ready to roll when, as if from nowhere, a naked Aboriginal man wearing only a hat wandered nonchalantly into the campsite as if the sight of four strange white men and an Aboriginal man aboard a massive, rumbling truck was an everyday occurrence. Blakeley wondered if this fellow might have some idea of the country into which they were heading and where they might find water, but no amount of creative sign language made any impact. The old man drew dog tracks on the ground and pointed south to show where the dingoes were. Blakeley understood: the man thought they were 'doggers'; professional dingo scalpers scouring remote regions with their native guides were not an unusual sight. Blakeley signalled that they intended to head west, which only seemed to alarm their new acquaintance, giving the expedition leader the feeling that this fellow might be an outcast, or at least not from these parts. Blakeley was sure he understood his gestures asking about finding water, and received the response that 'to take or interfere with water further west was likely to cause ill feeling'. Mickey tried to communicate further but only came away feeling that their visitor was frightened of him. Sutherland gave the man the name 'Rip Van Winkle' because he looked something like the fairytale character who had woken up from a 40-year sleep.

'I gave up trying to understand him,' wrote Blakeley, regarding it as wasted time. They threw him some food – an old damper Lasseter had made – and the five remaining expeditionaries, Blakeley, Lasseter, Taylor, Sutherland and Mickey, settled in for the 'Big Push' towards the border of Western Australia. But it would be only 200 yards before the matting was needed again.

The day turned out like all the others: hacking a way through the mulga, digging the truck out, laying out the matting.

Through the day, Rip Van Winkle and his wife and children, who had also seemed to appear from nowhere, followed them. They would regularly ask for 'ninjie' – but it was only when the wife made a pile of small sticks and performed the action of lighting it as if with a match that everyone understood. They must have had some contact with the Mackay expedition when they were here a few months ago. They gave them a box of matches, and that night they could see the smoke from a hidden fire nearby. Clearly Rip Van Winkle and his family were not in their own country. They were as uncomfortable and nervous about being in the area as everybody else.

There was a feeling of anxiety running through the camp that the expedition was being observed. Every morning smoke signals arose as soon as they began moving. According to Blakeley, the 'smoke experts can send up a dotted line of smoke that can stand up thousands of feet in the sky. It would be quite possible to see these dotted lines a hundred miles away . . . in the main they were only tick-tacking to others about our direction.' Nevertheless it added a level of uneasiness for everyone – particularly Mickey.

•

Every night, Lasseter would be seen retiring to the Thornycroft's cab, the inside of which was lit up with an electric light, and would dutifully record his thoughts and experiences of the day in his diary. It was a mystery to everyone as to exactly what he was writing about. Lasseter was always secretive to the point of paranoia. Blakeley remembered the diary as an impressive book, 'a nice-looking volume about eight inches long, six inches wide and perhaps two inches thick. It was well bound, with a very fine black morocco cover and exceptionally good paper.'

This particular night, Lasseter was unusually keen for a one-on-one chat with Blakeley, inviting him into the Thornycroft's cabin – 'This was his bedroom,' Blakeley recalled. Calmly, Lasseter made a momentous announcement. While Blakeley had been away searching for Coote, Lasseter said he had been working through various estimations of the gold reef's location and 'by my calculations we are approaching my country'. For the first time since the expedition started, Lasseter began volunteering information. He said that according to the bearings Harding the surveyor had taken when they were both out here years ago, the reef was about 100 miles ahead. He was certain he would be able to recognise something soon. There was a salt lake near to where they took their bearings. Blakeley asked him whether they were standing on the reef when they took these bearings, to which Lasseter said, 'No. We were some distance away. Where we took our bearings there was plenty of water. We had been hard pushed for water when on the reef country, so were not able to stay and only just examined a portion of it and took some samples, then had to make all speed back to the water.'

Blakeley asked him how far the reef was from the water and in which direction, but Lasseter was uncertain; he reckoned they were away from the water for about three days all up – and it took them about a day to get back.

Lasseter unfolded a map showing Lake Macdonald, and with a compass and ruler the pair calculated that he and Harding had travelled a distance of over a hundred miles. When asked where he took his reading from he replied, 'From a very high hill.' *A very high hill* . . . Blakeley scanned the map to discover Mount Marjorie about 30 miles east of the lake, at which Lasseter nodded, saying that could be it. Mount Marjorie and Lake Macdonald were squarely in line with where they were already heading, along the Tropic of Capricorn. Blakeley sat in shock. 'I was greatly relieved,' he wrote, 'for this was the first definite statement he had made since we started out.'

'Well, Harry,' sighed Blakeley, 'that's something definite.' Lasseter responded by saying it was the first time they had been able to talk without Colson being around. He didn't trust him.

Blakeley, quietly startled by this revelation, then asked Lasseter if he would give him the reading of the bearing, but the old prospector considered he had given out enough information for one day and closed the discussion. He told Blakeley he would have to trust him a little longer.

Lasseter's demeanour that evening reminded Blakeley of when they first met; of those preliminary meetings they had had in Sydney's Hyde Park when the discussions were open and trusting. Then, abruptly changing the course of the conversation, Lasseter opened his diary and began to read an entry he had written about Blakeley ('He was very severe on me,' recalled Blakeley) but then he turned the page and read another showing that he had found he was completely wrong and had changed his opinion. Yet Lasseter's newfound open-door policy in discussing the reef's whereabouts and his diary-read, sermon-like opinions of the expedition's members made Blakeley feel somewhat uneasy. 'It gave me a new angle on the man and confirmed my suspicions he was a newcomer to this part of the country.' Lasseter continued to read to Blakeley passages from the morocco-bound diary about what had gone on between the members of the party.

'Boiled down, it showed how suspicious he was of everyone,' Blakeley wrote. 'All through his writing it was plain that he trusted no-one, not even the Sydney crowd, for it showed that he feared once he revealed his reef, they would take charge, and not let him have a say in anything, although all that was stipulated in the agreement.' Further, Lasseter's writing revealed he wanted to have more say in regard to his reef and would therefore dictate to the company on his terms. 'The diary gave me some idea of what this little fellow was thinking,' Blakeley recalled. Lasseter informed him that when they found the reef, he would peg it in his name, and not that of the company. 'I did not argue over this,' remembered Blakeley, 'but knew that if we found the reef it would be pegged for the company, no matter how he read our agreement.'

What would be the repercussions when AWU boss Jack 'Ballot Box' Bailey back in Sydney learned that Lasseter had decided to put the reef in his own name? It was too frightening to contemplate. Blakeley mulled over the possible scenarios of what would happen when they actually did discover the reef – and what could happen if Lasseter pegged it for himself and Blakeley pegged it for the company. Men had been murdered out in the desert for less. Lasseter had already demonstrated he was not afraid of starting a brawl with Blakeley, while Lasseter and Coote had been in a scrap that could well have seen someone shot dead. There was no shortage of guns on this trip. 'I actually found myself worrying about it,' remembered Blakeley of Lasseter's intention to claim the reef for himself. 'For it was plain that Harry would put up a fight.'

23

WHO WAS LASSETER?

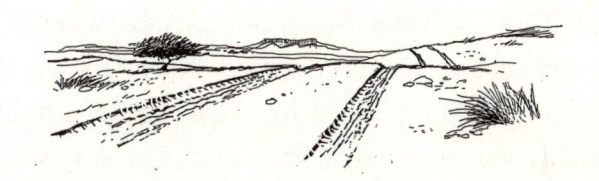

It is remarkable that so many people – intelligent people – had placed so much faith in and were even risking their lives on the basis of Lasseter's extraordinary story. Yet here they all were, hanging on his every word, hundreds of miles from any assistance, crashing through the unknown in the Thornycroft truck, headed for a destination to which there were no roads, no tracks, no coordinates, and so far no landmarks to give any clue as to where they would end up.

And, with Lasseter's behaviour becoming increasingly odd and fractious, it began to dawn on the expedition members that they actually knew very little about the prospector at all. Who was this brusque little fellow who had convinced everyone to charge headlong into the Never Never on the thread of a memory from 30 years ago? If only they knew. Harold Bell Lasseter was indeed a fascinating man.

Precisely who he was and what he actually did may never be fully understood. His life was a jumble of conflicting dates that jar with many of his extraordinary claims. He is hard to pin down, living as he did in a time long before the inescapable grip of the internet, long before digitised record keeping, his particulars chronicled in longhand, or typed methodically on mechanical ribbon typewriters and buried away in paper files in vast archives. However, what we do know of him only fuels his mystique.

Lewis Hubert Lasseter – as was his original name – was born in the tiny Victorian country town of Bamganie, near Meredith, between Ballarat and Geelong, on 27 September 1880. His parents were English. His mother Agnes died of sunstroke when he was very young, his father William John Lasseter remarrying. Lewis had three siblings: an older brother Arthur, who would be killed in the Boer War, an older sister Lillian and a younger brother Claude, who died at the age of nine from typhoid. His father was a seasonal shearer and occasional game hunter who supplied the Ballarat markets with rabbits, wild ducks and plains turkeys. He was renowned as a heavy drinker and was particularly harsh on young Lewis, frequently telling others his son was useless. A neighbour remembered meeting Lewis as a boy waiting with his father at the railway station at Meredith, and described him as 'a wee lad – wild – but not bad'. The same neighbour said it would have been about 24 years later when he next saw him in 1911 – 'as he'd run away from "where there is no place like it"', an adolescent gaol.

In 1896, 15-year-old Lewis Lasseter was arrested and placed in the local lock-up in Colac, Victoria, charged with burglary. He had been caught along with 25-year-old former shop owner Henry Boreham, the pair having broken into a local emporium called Belyea and Fulton's Economic Store and made off with a variety of goods. It was serious stuff, as the *Colac Herald* reported: 'The burglary had been perpetrated with great deliberation and care ... Lassiter [sic] had a mask, and both prisoners carried loaded revolvers; no doubt if they were discovered in the act they were prepared to take life.'

Young Lasseter had run away from home many years before, at the age of nine, and the well-respected Colac resident George Francis Sydenham and his wife became his guardians and employer. Sydenham had stumbled upon the stolen goods and revolvers in Lasseter's kit and handed it to police, who questioned the boy as to what he knew about the stolen goods. Sydenham said to him, 'It would be better to admit it,' to which Lasseter replied, 'Admit what? I know nothing about it.'

In the Geelong County Court, the ageing Irish judge Arthur Wolfe Chomley took a rather dim view of all this, handing down a sentence of six months' hard labour to Boreham and ordering Lasseter be detained in the Boys' Reformatory in Pakenham for two years. Of all Lasseter's remarkable experiences, this was one he never mentioned.

Lasseter's incarceration would prove ineffectual. Within a year, in 1897, he would escape from the reformatory in country Victoria, never to return.

From here on, his life story became a well-worn yarn thousands of Australians would come to know. The same year, the young absconder would be fortuitously found by an Afghan cameleer 400 miles west of Alice Springs, dehydrated, sun-bleached and clinging to life – clutching an oatmeal bag filled with gold nuggets.

According to another story, he served in the Royal Navy, sailing somewhere in the crashing seas of the Atlantic. Lasseter maintained that he had served for four years as a gunner in the Royal Naval Reserve aboard HMS *Powerful*, a heavily armoured four-funnelled cruiser. He said he was officially discharged from the service in 1901 – a claim he made when he attempted to join up for World War I, then saying his discharge papers were lost. If he left the navy in 1901, then that made the year of his enlistment

1897, the same year he discovered the gold reef. But on the odd occasion he pencilled on documents the date of his discovery as 1894, making him only 14 years of age when he had made his find. But then again, he had also said that in 1897 he was prepared to give up a hard life of seafaring in the waters around Cairns when he decided to look for rubies in the McDonnell Ranges. A lot happened during that year. Prospecting, naval service, seafaring – it was all very confusing.

However, he certainly did travel overseas. Around 1901, Lasseter sailed for England, and according to his writings spent some time visiting the Continent. He later claimed he had worked at the Harland and Wolff shipyard in Belfast and at various military arsenals, including Woolwich; apparently he even visited the giant Krupp armaments factory in Germany.

Sometime after this, he travelled to the United States where, in 1903, he married Florence Elizabeth Scott at Clifton Springs, New York, with a honeymoon at Niagara Falls. Lewis and Florence's daughter Lillian was born in June 1905, but a son, Arthur, born in 1908, died after only nine days. In his time in the United States, Lasseter became a practising Mormon, and for the rest of his life he would often spontaneously sing from a limited repertoire of American hymns.

While in America he enrolled in several correspondence schools, acquiring certificates in surveying, navigation and agriculture. In his 1989 book *Lasseter, in Quest of Gold*, author Billy Marshall-Stoneking notes: 'At least one of these certificates appears to be written in Lasseter's own handwriting.'

After Florence's mother died in 1908, Lewis became restless and persuaded his wife to sell the property in America and move to Australia. A neighbour remembered the newly returned Lasseter speaking with a particularly 'affected Yankee accent' – no doubt something intriguing to ordinary Australians.

They leased a 10-acre farm near the rural village of Tabulam on the Clarence River in New South Wales, about 500 miles north of Sydney, where Lasseter agisted horses 'destined for the Indian Army', the famous Walers as later used by the Australian Light Horse. He was also able to exercise his flair for writing as the Tabulam correspondent in the *Tenterfield Star* with his light column 'Tabulam Tinklings'. Lasseter was a particularly capable individual, building his own home, and piloting a small ship along the Clarence, occasionally ferrying cattle from Ballina to Sydney.

With the birth of another daughter, Beulah, the Lasseter family moved from Tabulam to Melbourne, where Lewis attempted to enlist in the army – a process that saw him rejected, then accepted, then discharged for medical reasons, whereupon he re-enlisted and was discharged again.

It was said that in 1918 Lasseter made another attempt to find his reef, taking five men from the railhead at Oodnadatta, heading west and then turning back due to a lack of water.

Certainly Lasseter was charismatic. In 1960 his eldest daughter Ruby wrote: 'My mother was continually hearing of his numerous infidelities – which he took little trouble to hide. For some reason women who had been quite respectable were fascinated by him.'

At about this time he met a nurse at the Caulfield Repatriation Hospital and, separating from his wife, married for a second time, without obtaining a divorce. It was a brief marriage, for in 1923 Lasseter met another nurse, Irene Lillywhite, apparently on a train, and married her the following year. Again, no divorce. They had three children, Robert, Betty and Joy, eventually moving to the Sydney suburb of Kogarah.

In 1924 Lewis Hubert Lasseter changed his name to Lewis Harold Bell Lasseter, a move that was probably inspired by a contemporary American pulp fiction author named Harold Bell Wright (1872–1944). A year earlier, Harold Bell Wright's novel

The Mine with the Iron Door had become a bestseller. The story was a kind of melodramatic contemporary western, where the characters have secret identities and furtive pasts, and, of course, there is a lost gold mine once operated by Jesuits in the 1700s, hidden within the *Canada del Oro* – the Canyon of Gold. Other books that may also have fuelled Lasseter's fascination with lost gold include Simpson Newland's *Blood Tracks of the Bush* (1900) and Conrad Sayce's *Golden Buckles* (1920), both stories dealing with the tantalising concept of an inland El Dorado.

Lasseter secured work on the construction of the new Parliament House being built in Canberra and the family moved to a camp for the project's labourers, perhaps something like the Hooverville shantytown built for workers on the Hoover Dam in the US, where they were free to build whatever they wanted 'so long as they put up something better than a tent'. During this time, Lasseter was able to further hone his journalistic skills, writing a satirical gossip column under the nom de plume 'The Gleaner'. He purportedly also had a wireless program, something well suited to someone who could speak so persuasively. In time, the family moved back to Kogarah Bay in Sydney, where Lasseter built the family home, with the concrete foundations ingeniously reinforced by rails discarded from the Sans Souci tramway line.

Lasseter would vociferously claim that he was in fact the person responsible for the concept of a single-span design for the Sydney Harbour Bridge. He was convinced that his idea was stolen. Interestingly, Lasseter did submit a detailed drawing of the structure as early as 1913, claiming he had put forward the plan in June of that year to 'the special committee appointed by the Government to decide on a suitable structure to span the harbour'. Lasseter's design is indeed similar but not identical to Chief Engineer John Bradfield's bridge, and certainly pre-dates it. In 1913 Bradfield was advocating for a cantilever-style bridge for

Sydney Harbour. In a 1929 letter published in *Australian Shipping and Steel*, Lasseter made his grievances clear: 'Seeing that I antedated Dr Bradfield by 10 years, I feel that I have more claim to be the original designer of an arch bridge for Sydney than he, and should at least be paid for the six months labour I spent on it, as I supplied working drawings and details down to the last number of rivets required. Yours faithfully, Lewis H. Lasseter.'

Ironically, back in Sydney he commenced as a construction worker building timber formwork on Bradfield's Harbour Bridge, but was eventually sacked for arguing about how the structure should be built. Money was forever a problem. Hugely in debt due to the construction of his house, Lasseter was desperate for work. In the slums of Redfern he took on managing a pottery workshop for limbless ex-soldiers.

•

Lasseter was a prolific letter writer, constantly penning his thoughts and ideas and firing off missives to whomever – in whatever government department – could make them happen. His letters are remarkably candid and conversational, addressing someone he had never met before in a way that suggested they had been part of some long, familiar conversation. In an age when letter writers would often close a request to a government official by referring to themselves as 'Your Obedient Servant', Lasseter would write casually to federal ministers with 'just a quick note' or wishing them 'fraternal congratulations' as though he too was a member of Cabinet. Certainly his brash and confident writing style was disarming to the recipient. His letters were clear, bold and amazingly authoritative on whatever subject he was peddling. When reading collections held in the National Archives of Lasseter's correspondence with officials, a pattern emerges. He would write to a relevant department with a proposal; the reply would generally be cordial and non-committal, to which Lasseter would respond, engaging in further discussion as if his idea was already in the process of worthy consideration. Correspondence would continue in earnest until eventually a frazzled departmental representative would scrawl a note urging to discontinue any further contact.

During World War I Lasseter wrote to the Minister for Defence with the idea of raising a 'Shotgun Brigade' for trench warfare, revealing, 'I once had an opportunity of witnessing the effectiveness of shotguns during a riot in a mining town in USA when 10 men with shotguns succeeded in routing about 200, variously armed with rifles and revolvers.' He enthusiastically outlined the worth of shotguns over machine guns and rifles, as a shotgun 'up to 50 yards will stop a man as effectively as a cannon-ball'. The minister's enthusiasm for the idea of a flying squad of trench-jumping shot-gunners wasn't quite on the same level as Lasseter's. Lasseter wasn't going to let the idea go cold, however, so he upped the ante. 'Kindly excuse my persistence in referring again to the matter of a shotgun brigade for trench fighting, my desire being to do as much as I can for my country . . . will you please advise me if you would accept a contingent of 100 men as an experiment providing I organise and equip them myself. Thanking you . . . Lewis H Lasseter, Tabulam, NSW.'

The minister's office responded by saying that 'such a unit would be thoroughly unsuitable' and further, the use of shotguns was 'a contravention of the laws of the Hague Convention'.

Hague Convention or not, Lasseter was on a roll, his next letter flagging that he was moving on.

'I regret that you have decided not to make use of a shotgun brigade, as I have on three separate occasions been a witness to its effectiveness when opposed to rifles, however it is of no use arguing the point. Will you please advise me if you would accept the offer of a bridge construction gang of either or either six or ten men, or if not, of myself individually. I am a bridge and structural engineer with a knowledge of surveying . . .'

One idea down, time for the next one. Surely, somewhere along the line someone would recognise his visions. Indeed, his proposal for a 'shotgun brigade' was not far from contemporary thinking: as it turned out, American forces during the Great War did employ shotguns, known colloquially as 'trench brooms'.

His mind leapt from one grand scheme to another. Another plan was for a long-distance mobile nine-inch artillery field gun capable of firing a distance of more than 100 miles. The barrel bolted together and could then be disassembled, and transported by a five-ton truck. Lasseter outlined that if the gun was given 'a fair trial', he should be paid 'a fair sum, say £5,000 . . . plus travelling expenses . . .'

The Department of Defence invited him to submit drawings or a model, but nothing more than his initial lightbulb of an idea materialised. 'Mr Lasseter has neglected to furnish detailed drawings or explanations necessary for the Department to arrive at a decision as to the merits of his ideas' was the department's response. 'Mr Lasseter has produced nothing but ideas . . . the onus is on him to develop his ideas and produce something concrete for investigation. Until this is done it is considered inadvisable to proceed further in this matter. Signed, Quartermaster-General.'

In 1915, Lasseter devised a proposal to overcome the stalemate that had derailed the campaign at Gallipoli in Turkey. He succeeded in securing a meeting with a Colonel Wilkinson, where he presented a plan for a Panama Canal-style excavation across the Gallipoli Peninsula, allowing the battleship the *Queen Elizabeth* clear passage into the Sea of Marmara, enabling the warship to lay siege to Constantinople. His idea was to use 50,000 men – under cover of a naval bombardment – to dig a trench north of the battlefields at the Bulair Isthmus, which connected the Aegean Sea with the Dardanelles. This would effectively sever the Gallipoli Peninsula from the mainland as an island. Despite polite rejections of the idea from the Department of Defence, Lasseter persisted, writing letter after letter until an exasperated official decided to shut the correspondence down: 'it would appear that you are unable to reconcile yourself to the non-acceptance of your suggestion. This is regretted, but the officers who advise are not without experience in these matters, and the execution of Military and Naval operations . . .'

Unfazed, Lasseter then volunteered the idea of a bullet for a smooth-bore gun. Another scheme proposed building a kind of grand industrial city for unemployed and incapacitated returned servicemen looking to retrain and re-join the workforce. He had found just the site for it at Corner Inlet in Gippsland, Victoria. The interminable correspondence about his unsolicited idea became too much, the Repatriation Commission turning him away. Handwritten notes at the edge of one particular letter state, 'The view of the Commission is that this is too nebulous . . . a communication should be sent to Mr Lassiter [*sic*] in diplomatic language pointing this out and at the same time couched in such terms as to discourage further correspondence, otherwise it is probable that it will involve the Department in a lot of unnecessary trouble that will lead to nowhere . . .'

Perhaps it is his military service during the Great War that reveals something more of Lasseter. He was 34 when the war broke out – hardly the typical age of most Australian volunteers. Nevertheless, he sought to enlist.

His army service file in the National Archive is particularly revealing. He attempted to enlist in Melbourne during February 1916, his application declaring he had served four years in the imperial Navy, leaving England in 1901. He put down his profession as a bridge engineer, no doubt aiming to join the engineering corps. His application, however, was rejected. He tried again two weeks later, this time applying in country Victoria, where he was passed, joining the 4/3 Pioneers at Seymour, but in October 1916 was discharged as medically unfit. In August 1917 Lasseter re-enlisted, his response to the question 'Have you ever been rejected as unfit for his Majesty's Service?' being 'No'. He was accepted, interestingly applying for the Flying Corps, where his application was approved and then annulled on the same day. He was brought before an army medical board on November 1917, and the medical assessment described Lasseter as 'mentally deficient . . . Has marked hallucinations, wants to join the Flying Corps as a friend is going to present him with an aeroplane . . . Has peculiar manner and is constantly talking . . .'

He was discharged.

24

NEW DIRECTIONS

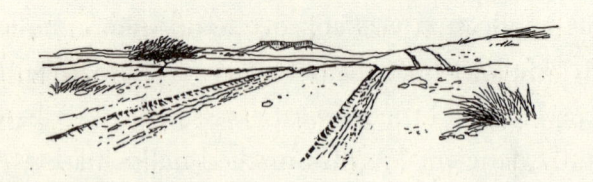

The Australian Inland Mission Hostel in Alice Springs had suddenly become the focus of attention as people came running from all over town to find out who was the bleeding casualty brought in on the back of Fred Colson's truck. Even the Sunday cricket match was abandoned. Everyone wanted to know what had happened. Word soon spread among those crowded around the hostel's gates that it was the pilot from the Lasseter expedition who had crashed his plane somewhere out past Haasts Bluff.

Coote had been fairly badly bashed up. The aircraft's joystick had broken two of his ribs, a piece of metal tubing had holed his elbow, and the sizeable cut to his leg had been caused by his impact with the compass mounting – the result of not wearing his safety belt. Colson reminded him of how much worse things could have been: 'Your head was jammed against your chest. It's a wonder you didn't break your neck.'

Coote probably wished he had. He was the Central Australian Gold Exploration expedition's official pilot. But now there was no more plane and therefore no more need for Errol Coote. He was stranded – bedridden in a hospital in Alice Springs, a place inconveniently located somewhere between Sydney and the expedition now crawling towards the Western Australian border. But Coote was determined he wasn't going to be left out of the picture – he would be in for the kill when they found Lasseter's reef, and he wasn't going down without a fight. As soon as he could, Coote started sending telegrams to Sydney suggesting the *Golden Quest* could be repaired, writing a list of what he thought it required. From his hospital bed, he then cranked up his journalistic skills, embroidering thrilling accounts of his adventures for the *Evening News*. A shop owner in Alice Springs named Ohlsen decided to cash in on Coote's growing publicity, also penning lurid tales of the plane crash, 'much of it being quite erroneous', wrote Coote. But he records a curious side note: 'The other fellow had sent something through about a reef in the Petermanns and the directors thought it was from me.' These articles were picked up by other newspapers as well, and word of the flamboyant stories soon got back to the directors in Sydney, who sent Coote a blunt three-word telegram to 'cease senseless chatter'.

Coote received a message from the Sydney office that a replacement aircraft was on its way to Alice Springs piloted by a young Sydneysider named Pat Hall, who like himself had been trained by the New South Wales Aero Club. Coote was being sidelined.

Hall in due course landed at the aerodrome in Alice Springs in another DH60 Moth biplane, nowhere near as flash as the *Golden Quest* had been when it started out. It was indeed well worn – to the point of its airworthiness being questionable. It had a smaller engine than the *Golden Quest* – a Mark II Cirrus, with a much shorter flight range. Hall brought with him terse instructions from the Sydney office that Coote should hand over all his maps to the young pilot and instruct him as to how to reach Ilpbilla to receive 'up-to-date reports' from the expedition. In effect, Coote had been relieved – if not sacked – as the official pilot. He needed time to figure out a way to keep his hand in the game.

Hall had clearly been rattled by his long solo flight from Sydney, particularly on the barren featureless stretch from Oodnadatta. Coote was quick to unsettle Hall a bit more. 'I laughed about the maps,' he recalled. 'There were two maps of the centre in town, both issued by the Commonwealth Department of Maps and Surveys, and both had prominent landmarks 40 miles out of position – depending, of course, on which map one accepted as being correct.'

Hall now looked even more anxious. 'A man would only be looking for trouble trying to find Ilpbilla. It is only a name on a map isn't it?' he asked Coote, who immediately saw his opportunity.

'Yes,' he replied, 'but if you wait a day or so I will take you out there myself. You can take the plane off and land it, and I will fly it out there. I know the way all right. I've flown over most of it already.'

'That'll do me,' sighed a relieved Hall. Bingo – Coote was back in the race.

•

Lasseter and Blakeley's discussion in the cab of the Thornycroft that night had been extraordinarily productive. Lasseter had confided to Blakeley that his 1897 bearings were taken from 'a very high hill' and both had agreed that Mount Marjorie was the logical target. It made sense: it was near a lake and it was in line with where they were travelling. The relationship between the leader and the guide was the best it had been since the expedition began. Lasseter had asked Blakeley not to tell the others about their conversation, so he just informed them that they would know within a week whether 'the gold reef was really there, as this was the country in which it was supposed to be'. For Sutherland and Taylor, this was some consolation at least. Driving an ailing truck even deeper into the wilderness with no clear objective had been causing them to worry considerably. If the going had been tough before, it was now growing worse. They were slowly travelling across a vast plain covered with big tussocks of spinifex, which would suddenly stymie the big truck when the cast-iron differential ploughed into it. They would have to dig it out from underneath with a shovel. 'It was a brute of a job,' recalled Blakeley. 'To give some idea of how slow it was, we travelled only six miles in four hours until the tussocks got thicker.'

Then something extraordinary happened. Lasseter suddenly announced that very soon the landscape would change from the worthless desert they had been battling and they would be in fine mineral-bearing country. And it happened, right where he said it would. 'It was one of the most extraordinary coincidences we encountered on this trip,' remembered Blakeley. 'As soon as Lasseter told me we were approaching where he thought his reef would be found, the character of the country changed completely.' Sutherland, the miner of the group, was astonished, finding quartz and ironstone lying all over the place. They stopped to set up camp, Sutherland and Blakeley dollying some stone but finding not much more than pyrites. That evening was the first time any serious discussion about mining had taken place since the expedition began. Things were starting to look promising. The race was on to get Lasseter to Mount Marjorie as soon as they could to see if this was where he and Harding the surveyor had taken their bearings.

But the journey to the mountain was heartbreaking, with the countryside impenetrable, 'all broken up and impossible to get a truck through', wrote Blakeley. Mount Marjorie would occasionally seem tantalisingly close, then, through their meanderings to negotiate the terrain, edge further away. In trying to negotiate their way closer, after a whole day's driving they were virtually back at the campsite from the night before. In the end, after exploring every conceivable way to approach the mountain with the truck, it was decided that the following morning Blakeley and Lasseter would walk through the scrub to climb it.

That night was unpleasant. Wind whipped up waves of stinging sand that forced them to rig a tarpaulin for cover. 'It was a rotten camp,' remembered Blakeley, 'sandy, full of thorns, windy.' But they could take comfort in the fact that tomorrow Lasseter would – Moses-like – climb the mountain, and triumphantly descend with something no doubt even greater than the Ten Commandments.

By 6.30 am Blakeley and Lasseter were pushing their way through the scrub, hiking through four miles of rough country to reach the base of Mount Marjorie. It was rough going, and when they reached the base of the 809-metre-high monolith, they simply couldn't find a way to even begin climbing it. Eventually they found a big crack in the rock and began shimmying up – Lasseter proving a remarkably adept climber – but it came to a dead end at an overhang and he couldn't proceed any further. 'I spent some very anxious minutes watching him climb down,' remembered Blakeley, who helped him by physically guiding his feet into footholds. Lasseter commented at the time that he was worried he might have fallen from the rock face and thanked Blakeley for his assistance. Blakeley noted his appreciation, as 'Harry very seldom thanked anybody and this lot of thanks surprised me'.

They walked for another mile around the mountain's base before finding a possible way up. The climb was gruelling and perilous, and loose scree gave way under their feet as they scrabbled to reach the top. But the vista from the summit was something else. 'We soon forgot our hard climb – the view was something I shall always remember,' wrote Blakeley. He could plainly see the faint purple silhouette of the Petermann Ranges on the horizon, and even though Lake Macdonald seemed so large and close, it was still 30 miles away, the sun's reflection on its salt crust almost blinding. Bushed, Blakeley sat down to wait for Lasseter to take the bearing. In a few moments they would have the greatest gold find in history in their grasp.

At a few minutes to noon, Lasseter began to prepare for the sextant reading; as always, there was much fussing – unfolding pieces of paper and organising his watch and his lead pencil, removing the sextant from its case and squaring his feet in the right position while he readied himself to 'take the shot'. After taking his bearing, he took up the pencil and proceeded to perform some calculations on paper. Blakeley waited while the old prospector finished what looked like some serious arithmetic. At last he had arrived at his conclusion. He turned to Blakeley and knowingly exclaimed, 'Well, I'm damned if these figures are not the same as those we left in the vault of the bank in Sydney.'

Blakeley's eyes widened. What did he just say? He wasn't sure if he had heard correctly. 'Do you mean Harding took his bearings from up here?' he cautiously asked.

'Oh no,' replied Lasseter dismissively. He said that he and Harding hadn't realised their watches were out by an hour and a quarter until they had left here and travelled all the way to Carnarvon some eight or nine hundred miles away on the Western Australian coast. Lasseter said that he needed to do some more arithmetic to find his bearings. Flustered and impatient, Blakeley wanted to dispense with all this navigational gobbledegook. He wanted Lasseter to tell him straight – what did he propose they do now?

'I want to go one hundred and fifty miles further south to pick up my bearings,' Lasseter said without batting an eye. Blakeley was incredulous. *One hundred and fifty miles further south?* Somehow, the expedition leader kept his now renowned temper under control. If they were 150 miles too far north of where the reef was, then why the hell were they standing on top of Mount Marjorie? Lasseter was only too happy to explain. He said he'd had 'everything figured out' in Sydney. 'I could have told you from the word go,' he informed Blakeley, 'that our bearings would be near Lake Macdonald, but I had to come here to put myself right.' Blakeley, as expedition leader, was faced with the unthinkable – this really was a wild goose chase. Everyone had put their lives on the line to follow Lasseter's whims, fighting their way across dangerous, trackless wasteland – an ordeal that had almost seen their pilot killed, written off an expensive aircraft, virtually destroyed a specialist heavy-duty truck given to them on loan, and wasted shareholders' money – so that Lasseter could come here to 'put himself right'.

Blakeley then asked, 'When you get there, is that where Harding took his bearing?' Where Lasseter was now proposing to go was over near the Petermann Ranges and Blakeley knew there was no mineral-bearing country out there. People had been looking for gold in that region since 1908, every time coming back broken-hearted and empty-handed. If Lasseter had said 150 miles north, perhaps that would be a different story, because the country up there could possibly contain gold. And that region tied in better with his stated times and distances from Carnarvon. Maybe he had his figures wrong. Lasseter checked his sums and then shook his head, adamant his calculations were correct. 'It won't work. It's south right enough.'

For Blakeley, the search for Lasseter's gold reef was as good as over. As far as he was concerned, the CAGE expedition had been duped, and as leader – and a key person in endorsing Lasseter's story – he felt a worrisome responsibility in having recommended the search to the board as feasible. He was, however, prepared to give Lasseter the opportunity to finally own up to his cock-and-bull story about a gold reef out here. 'Now look here, Harry,' Blakeley said, 'where did you get this tale from? Tell me everything, for the direction you want to now go is right into the big breakaway country. So far even camels have never crossed that country.' If Lasseter had ever set foot out there, he would have known it.

Lasseter had suddenly lost control of the whole situation. He knew Blakeley's patience had finally come to an end and that he was ready to abandon the expedition and report to the board in Sydney. He gave it one last shot to save his story, and himself.

'I have known for some time that you have doubted my story but I want you to play me fair,' he confided to Blakeley. 'I have no complaints up to date of your treatment. We have had our differences and some exchanges, but you are like myself, quick to forget, and I beg of you not to go into a lengthy statement of my story. I do admit some things were said to mislead you. Why I did that was because another member of the party has made me suspicious so I told little bits to mislead you, and I want you to trust me a little longer.'

Blakeley sat silent and unmoved. Lasseter needed to keep his plea alive, asking him how many weeks it would be until the hot weather. Blakeley replied about five or six at the most. 'Well, give me that much time, for I am quite sure I will have enough indication by then to convince you this is not a dream yarn. If the country south of here is as you say, then I would think there is something in your idea that I am working out these figures wrongly.'

Lasseter looked for any sign Blakeley might waver. The expedition leader had heard his appeal and was grimly considering what his next move should be. Lasseter had confessed that he had lied to Blakeley – he had lied to everyone in fact – and yet he was begging for Blakeley's trust. What was to say the whole gold reef story was a lie? But then that exasperating recurring question inveigled its way into Blakeley's thoughts: *what if it was true?* What if the reef did exist? They had come this far, and spent so much time and shareholders' money blundering around out here – what was there to lose?

Blakeley recalled his predicament. 'After a long talk he asked me to give him a second chance.' He spelled out to Lasseter precisely what he thought. 'If you did the honest thing you would return and tell them that you failed to give me one single incident that was convincing. Your story does not knit up together.'

The message was unequivocal, but somewhere in the tone of Blakeley's voice Lasseter detected he could still be in with a chance. He unfolded the map, telling Blakeley, 'I did not intend to give you this information yet, but you see those hills marked the "Three Sisters"? Well, that's where I want to go for I am sure I can relocate myself once I get there.'

Blakeley took a deep breath. It would take a lot to convince the expedition's leader of the wisdom of gambling with Lasseter one more time.

•

The descent from Mount Marjorie had been every bit as dangerous as the climb, with Blakeley and Lasseter thoroughly exhausted on their return to camp at dusk. Seeing their approach, Taylor met them with a waterbag. It had been a long, unrewarding and pointless day. After dinner, Lasseter retired to his usual place in the Thornycroft's cab, and under the electric light began writing up the day's events.

'I had a bit of a feed and turned in,' Blakeley recalled, 'for I was greatly troubled. The show today had been that of a simple child. I could not pin Harry down to anything definite. I lay there and thought.' The more he thought, the more the things Lasseter had said weren't adding up. For someone who had supposedly been here twice, Lasseter didn't seem to recognise any of the unmistakable, unforgettable landmarks they had seen. When quizzed on why he couldn't remember the landscape through which he had travelled, his answers were more or less the same, saying, 'I was so excited at relocating the reef, I didn't pay much attention to the country.' His story about travelling from Carnarvon and back with Harding the surveyor was riddled with so many inconsistencies it seemed more implausible every time Blakeley ran it through his mind. All of these dubious aspects would be canvassed in his report to Sydney.

Then, as Blakeley was on the edge of sleep, Lasseter called out to him, asking whether he remembered the figures of the bearing he had taken with the sextant on top of Mount Marjorie. Blakeley reminded him that he had written the figures down in his book. Lasseter replied that he must have lost that page: 'I can't find it, that's a darned nuisance, for it means I shall have to go out again tomorrow and take the reading again.'

Blakeley was incensed, curtly retorting, 'That's stiff luck.' It was not lost on him that when the bearing was taken on the mountain top, Lasseter had exclaimed that he was 'damned if these figures are not the same as those we left in the vault of the bank in Sydney'. That Lasseter now said he had no idea what those figures were infuriated Blakeley. 'He could remember that and yet asked me such a foolish question,' he recalled. The expedition leader was a hair's breadth from declaring the whole enterprise finished. 'I thought, "This will be sufficient to end the expedition," but realised I would have to wait and see what he did the next day. He was evidently sparring for time.'

The following morning, Lasseter approached Blakeley to see if he was ready to go and retake the bearing, to which the expedition leader declined, suggesting Taylor and Mickey might like to go instead. To Blakeley's disbelief, Lasseter decided he wouldn't bother climbing Mount Marjorie again – instead: 'I'll go on top of that hill over there . . .'

While they were gone, Blakeley confided his concerns about Lasseter to Sutherland, the wisest sounding-board on the expedition. Blakeley was confused as to why Lasseter needed to perpetually climb things to ascertain his location with the sextant. 'Sea captains can take it from the deck of a ship,' he breathed: Lasseter only needed an uninterrupted view of the horizon. There were lots of little things like this that didn't make sense. Sutherland understood his reservations, and related a discussion he'd had with Lasseter when he simply asked him how he'd managed to travel out to Cloncurry on his journey out to the reef back in 1897.

'Oh,' Lasseter replied, 'I had the mining fever, so I thought I would hop in a train and have a look at the new copper field.'

'You went out by train?' asked Sutherland carefully.

'Yes, and a pretty rough trip it was too,' Lasseter recounted. 'But things were all excitement there over the new ruby find, so I thought I would go to the ruby fields.'

Sutherland didn't comment on Lasseter's train ride to Cloncurry in 1897. As someone who had actually worked on the railway's construction until the line was opened in 1906, he didn't need to. The conversation that then unfolded between Blakeley and Sutherland began to unearth all sorts of inconsistencies in

Lasseter's stories. The incident when he was found dying in the desert by a wandering Afghan cameleer had at various times been reshaped with Lasseter being rescued by a dingo scalper or even Harding, the mysterious surveyor himself.

But the fact was that Lasseter had confessed to the leader of the expedition that he had been lying as to the whereabouts of the reef. Blakeley and Sutherland thrashed out the one key question that afternoon. Should the whole thing be called off? Blakeley believed that if the expedition was cancelled due to Lasseter obfuscating the truth, then his explanation to Sydney would be that he did so because he was convinced Blakeley and Colson were colluding to double-cross the company. In this hotbed of brewing paranoia, Blakeley knew who Jack Bailey and his mates would believe. What to do? In the end, it was decided that the wisest course of action would be to continue with Lasseter's 150-mile rethink.

When Lasseter returned from his second bearing reading that afternoon, he proclaimed that he was never more confident that his decision to head south was correct. Blakeley asked him if there was any difference between the bearing taken today and yesterday. Practically none, Lasseter said.

The decision was made. They were heading south. They just needed to tell Colson. They made a bonfire to send a big smoke signal. At night a brilliant shaft of light from the hand-held electric spotlight scoured the night sky, able to be seen from miles away.

•

The Thornycroft's suspension was suffering badly. Weeks of relentless, punishing treatment saw a front main leaf spring break, a repair that couldn't be ignored. The truck had already experienced cracks in the rear springs, but this break at the front would ultimately create serious problems for the steering. Taylor once again proved his ingenuity by jacking up the back of the truck and removing a broken rear main spring, customising it to fit the front. They were ready to move.

Whatever direction the expedition needed to take to comply with Lasseter's new destination, the options were bleak. Blakeley wondered if they could follow the most direct route, travelling through the salt-lake country near Lake Macdonald. The next morning at daybreak, the truck set off on a reconnaissance journey to see what the flat salt-country beyond the sandhills was like. Not wanting to risk taking the Thornycroft too far, Blakeley decided to continue on foot, and discovered the land had become a brittle crust: 'I only walked a few feet on it and decided it was impossible for the truck. It was soft soda on salt ground, anything could be under it.' He felt uneasy. Miles in the distance he could see very rough territory. It reminded him of the terrible breakaway country he had spoken of to Lasseter. Blakeley returned to the rest of the party, telling them it was impossible to get through. Their only option was to turn around, head back to Ilpbilla and make south from there.

•

The return drive to Ilpbilla was miserable. A blistering wind saw the Thornycroft's radiator boiling to the point where it used 33 gallons of water for the day. On reaching the camp that night, Mickey asked Blakeley if he could sleep underneath the truck. He was clearly scared. Blakeley assured him there was nothing to worry about, but Mickey had been nervously observing the increase in smoke signals. They had been intruding into territory in which they didn't belong. Taylor had been watching this intense conversation unfold and asked Blakeley what their Aboriginal guide was worried about. 'I told him he had the wind up and expected a lot of natives to arrive,' recalled Blakeley. 'Phil was quite concerned and wondered if it was wise for us to be so far away, as we were sleeping a hundred feet or more from the truck.' This issue of safety at night was always a dilemma – a campfire near so much fuel could prove even more catastrophic.

The sun's rise the following morning took away a lot of the angst about a sudden attack from wrathful Aborigines. For breakfast they ate leftover stewed pigeon from last night's meal – birds Taylor had shot when they had arrived. Lasseter approached Blakeley to inform him that he wanted to explore a valley on the other side of a mountain range, and the expedition leader told him to take Mickey with him. But Mickey soon returned alone, too frightened to cross the foothills. Blakeley, worried their only hope in locating the gold reef was now wandering the wilderness on his own, questioned Mickey as to why he had abandoned Lasseter. Blakeley recalled, 'He said, "Too many wild blackfellow plenty make sign talk about us." I saw he was badly scared. White people often get their first glimpse of real terror when they see a native's fear of his enemy brethren.' But Mickey reassured Blakeley that Lasseter would return, as he had been marking his path with white ochre. The old prospector eventually came back telling of how he had walked over the range about 10 miles to explore a big valley – but Blakeley thought he didn't look nearly tired enough to have completed such a lengthy round trip. Suspicious, Blakeley took Mickey with him the following morning to see how far Lasseter's chalk-marks extended, and at about five miles they ended at a clump of trees 'where he had evidently put in the day and had a good rest', remembered Blakeley. 'That to me, was pretty convincing proof that Lasseter had not had that much bush experience.'

It was clear that despite reluctantly agreeing to follow Lasseter's new path, the expedition leader had lost all faith in Lasseter and the veracity of his gold reef. Blakeley was in an invidious position. If he pulled the pin on the expedition now, then he alone would be held responsible for its failure – held up as incompetent, unprofessional and unfit for his role. The galling result would be that Lasseter would survive unscathed – both he and his farcical fairytale intact. What needed to happen was for Lasseter to be exposed for the fraud he was. Blakeley from now on would comply with whatever was required to search for the reef, but he would be watching Lasseter like a hawk. He needed to get hold of hard evidence so that he could call it quits on this charade and emerge with his reputation still intact. He was going to give Lasseter enough rope to hang himself.

The next day that looked like it just might happen. As a result of the fire on board the Thornycroft, it was agreed the truck should never be left unattended for long. Any number of things could happen that could see it ransacked or burned to the ground. Blakeley agreed to guard the truck the next day when he spied an opportunity he had never had before. Lasseter had left his big tin trunk unpadlocked. The question was – should he open it? Blakeley's curiosity got the better of him: 'I did the meanest thing I have ever done in my life – I opened Harry's tin trunk and went through all his papers, also the box with some gold specimens.'

Blakeley was suspicious that somewhere along the line these gold stones would be miraculously produced from Lasseter's pocket after he returned from a long walk, claiming he had rediscovered the reef. Blakeley wrote that there were seven gold specimens, and from Taylor's toolbox he took out a specialist breast-drill; pressing down on it with his chest, he drilled miniscule holes in the gold stones. He then poured white metal into the holes, concealing them with glue. 'I did this so that if ever the stones turned up again I could identify them.' Blakeley's excuse for rifling through Lasseter's trunk was that he was looking for Blakiston-Houston's navigational almanac, but he found more than that. There were all sorts of other curious items inside, including an American passport with a different name. 'I found plenty of food for thought in that trunk,' he wrote. 'Seeing that there was such important evidence in it, it's a wonder he went away and left it unlocked.' Yet it seemed there was no 'smoking gun' find among Lasseter's belongings that could prove conclusively he was undertaking some kind of fraud.

When Lasseter returned at dusk, Blakeley noticed that the first thing he did was open his trunk. During the evening, Blakeley went over to it and found it was padlocked again, and concluded that 'leaving it open was evidently an accident'.

Long, hot days at the old airstrip were spent waiting for the return of Colson in the Chev truck, with Blakeley determined the expedition would not turn a wheel until he got there: 'I told them I would not go on again until I knew for sure the auxiliary truck was coming, for if anything happened to our turnout our existence would depend on being relieved by it.'

And rightly so. They had already experienced a terrifyingly close call when the Thornycroft caught alight on the drive to Ilpbilla – the memory of it never far from everyone's minds. The truck was a rolling mixture of fuel, electricity and explosives. 'The fire danger was very real, more especially now that warm weather had come in. We always carried the fire extinguisher on our knees when travelling. We packed drums of water and provisions right at the back in case the fire got out of hand. That is why I was so anxious to find out about Freddy. Should a fire get out of hand we might be able to pull the water drums off but we carried a lot of petrol besides the explosives, also the gas cylinders of the oxy-welding plant. A good place to keep away from once a fire got out of hand.'

The Aborigine Sutherland had named Rip Van Winkle reappeared, this time with a bedraggled group of three women and six children, asking for more matches, and they were given some more leftover food. Four of the six children had serious deformities. Then a tall, bald Aboriginal male appeared as part of the group. Blakeley wrote that they tried to communicate: 'We could not make any sense out of them at all, but it was quite plain that they had contacted whites before.' But 'Old Rip', as he was now being called, and his bald friend would continually look out west nervously, as did the women. Mickey told Blakeley they were frightened of the increasing number of smoke signals, and that pretty soon 'plenty more blackfella come along'. It was just a matter of when.

Early the following morning, it happened. Someone in the camp observed that they were being watched from the bush; they could see a face in the mottled shadow. It was what they had been dreading. Everyone knew it would be one of the unseen Aborigines who had been sending up smoke signals during the previous days. Blakeley told the others to stay while he slowly walked halfway towards his visitor, before giving a signal to sit down. To his surprise, the stranger turned briefly and began talking to others. There were more, but just how many more? Blakeley attempted to communicate with simple signals he had learned from a tribe from nowhere near here, wondering if these signs were universally recognised. He held up both hands and lifted each foot to encourage the stranger to walk across and meet him halfway. To Blakeley's delight he understood and walked over. The pair sat down, and the Aborigine talked for some time, pointing to the west and south-west, and when Blakeley made the sign for water he readily understood, indicating to him where the waterhole was. Blakeley began drawing maps in the sand indicating where they were proposing to go, when four more Aboriginal men and two small boys appeared as if from nowhere.

Blakeley had never seen Aborigines like them, markedly different from the tribesmen he had encountered throughout the outback. They were tall and light-skinned, Blakeley describing them as only a shade darker than his sunburned arms. They wore a charcoal marking across their chests, 'the charcoal showing up quite distinctly' on their light skin. Blakeley observed, 'Mick in contrast to them, looked really black.' Blakeley would write of these Aborigines in awe, 'These men were fine to look at . . . of very fine physique and well proportioned . . . their bodies clean, their hair was free and well kept . . . If science had the fashioning of man it would use these men as a pattern.'

There was no indication these people had ever come into contact with white men before.

The small boys were fascinated with the truck, running their fingers along the tyre treads, gradually realising these were what had made the curious tracks in the sand they had been following. Taylor opened the truck's bonnet to show them the engine but they were more intrigued by the operation of the hinge. Sutherland gave one of the boys a pocket mirror, and when he looked into it and saw his reflection he laughed, only to be surprised that when he turned it over there was nothing on the other side. He flipped it over and once again burst into laughter. To the expedition members' delight, the small boy held the mirror sideways to his eye, trying to keep one eye on both the reflection and on the back at the same time, wanting to know how the laughing boy he saw could disappear so quickly.

Convinced that this was their first contact with whites, the morning became a sort of exciting show-and-tell for the expeditioners, keen to impress on their stone-age visitors the wonders of what they had brought to the desert. There was no shortage of theatrics. 'Phil got out some petrol and poured it over the sand, then acted the fool like a magician does, passing his hands over it; then he lit it,' Blakeley recalled. A huge sheet of flame shot up spectacularly, but the effect on their audience was not as hot. As Blakeley described, their expression 'changed to one of surliness and they kept muttering as though they were annoyed; in fact I thought they were going to clear out, so I kicked some sand over the fire'.

At lunch the expedition members made bread, butter and meat sandwiches. 'They hadn't the faintest idea what to do with them,' remembered Blakeley, 'clear proof that this was their first handout. I had to take a bite out of the leader's sandwich and show him what it was; soon they were all eating.'

But it was water boiling in the billy that intrigued them most; one of the men dipped his finger and scalded himself. 'Of course it burnt him,' wrote Blakeley, 'and you should have seen the ugly scowls on their faces. I saw at once this was a mistake and these men would resent any tricks being played on them so I put my finger into it quickly, then into some cold water. I did it two or three times. Then the young fellow with the burnt finger put it into the cold water. That was all right . . .'

At two o'clock in the afternoon the fiercest component of the day's heat was about to set in. Exactly what they were going to do about Colson – whether they should set out on a search mission – was a decision they would need to make very soon. Every moment wasted brought the impending wrath of summer closer. In the meantime all they could do was stay put and kill time, waiting alongside the two groups of very different Aborigines at the old aerodrome. The skyline possessed a fierce shimmer, melding the sky with the land, and only those with the keenest eyesight noticed a miniscule aberration in the distance.

One by one the Aborigines slowly stood up to look at the horizon. There was considerable conversation. They had seen something out there, but exactly what it was, no-one could be sure. Far in the distance, tiny shapes were shifting, evolving as silhouettes through the simmering haze to reveal a string of camels plodding towards them, and now the outlines of three men riding on top came into view. The Aborigines seemed particularly distressed, pointing and apparently arguing with each other as the camel train trudged slowly towards the camp.

Lasseter, who had until now been sitting on a benzene tin in the shade of the bough shed, stood up and wandered out into the heat, squinting into the haze disapprovingly. 'This is one of the parties looking for the reef,' he said grimly.

25

NEWS

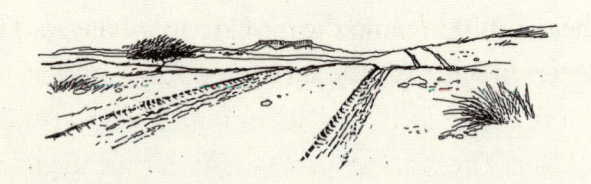

Five camels ambled into the middle of the camp with one white man and two Aborigines in stockmen's gear riding the first three, the remaining camels carrying supplies. If they were a mining party in search of the reef, they certainly weren't well equipped; they were travelling far too light for such an expedition. The white man, who seemed remarkably young – in his very early twenties perhaps – dismounted from the braying camel, handing the reins to one of his native men, and asked the gathered group who was in charge. Blakeley wandered over and introduced himself. The young man spoke to him with a distinct European accent. 'I was surprised to see fresh tracks up at the waterhole, so I came along to see what all these truck-marks meant,' he said to Blakeley in a startlingly brusque manner. 'I'm pretty short of tucker and thought you might be able to let me have some.'

Blakeley wanted to know what he was doing out here. It seemed more than a coincidence to find a young white foreigner with two natives bumping into a gold exploration mission in the middle of absolutely nowhere. Blakeley was sure something was up – as were the others – and began to question the three new arrivals while he boiled the billy. The young man introduced himself as Paul Johns, a 22-year-old German who had arrived in Australia about three years earlier. He had taken up work outback as a dogger, and had just travelled from the Lutheran mission at Hermannsburg, bringing with him the two Aboriginal boys as guides.

Blakeley was very familiar with doggers, who would travel into the most remote and lonely regions looking to skin wild dogs for pelts. But the expedition had seen no evidence of dingoes at Ilpbilla. They were questioned as to whether they knew of any other mining parties roaming the district, to which Johns said he wasn't aware of any – they were only here for water. Still, Blakeley wasn't comfortable with the young dog hunter being around.

In 1930 the dingo was considered a vicious, destructive pest, attributed with killing livestock across Australia. Known as a 'native dog', it would be eradicated by farmers whenever they had the opportunity. But the wandering mercenary the professional dogger was often regarded with disdain. Of all the tough and lonely livelihoods found in the outback – of hard cattlemen, of drovers, of miners – the work of a dogger would, for some bushmen, be resoundingly reviled. These men didn't dig for minerals or raise cattle or crops; nor did they muster livestock on the thousand-mile-long stock routes that crossed the top end of the country. They were professional killers who would stock up on ammunition, take a selection of boning and skinning knives, a few bags of phosphorus and strychnine, and head out bush to poison and shoot dingoes for the government bounty placed on their pelts. They told pub mates they poisoned the dogs with 'strychnine sandwiches'. There was none of the sport of hunting in this.

Doggers were often employed on cattle stations primarily as rabbit poisoners, but on occasion, like Johns, they operated independently, purely as bounty hunters, skinning dingoes and bringing the carcasses into town, presenting them to a shop owner appointed as a representative of the Marsupial Board. A bounty would only be paid to the dogger providing the scalp was still joined to the tail by the full length of the skin of its back.

While visiting Australia, the British explorer Walter Kilroy Harris gave a detailed account of his experience as to how the dogger performed his gruesome and repellent work of poisoning and skinning. 'The dogger places a piece of fresh meat with strychnine rubbed well in. The dingo prowling round at night will not be long in getting to the scent, and will follow it up until he comes to the first "bait". After devouring the poisoned meat, certain internal sensations will make him set out for the nearest waterhole as fast as his legs can carry him. Very likely he will die before he gets there, as the doggers are careful to lay the baits as far as possible from water. Or he may have a fit after drinking, recover, travel until his strength has gone, and then have another fit. I have never known a dog to live long after the second fit.'

Harris's opinion of the outback dogger was that these men were hardened and desensitised, unperturbed by the condition of the dingoes they skinned. 'If a dogger finds a dead dingo in a waterhole it is no use to merely report the find'; the dogger would have no qualms in hauling a rotting carcass from its watery grave to collect his bounty. Otherwise, as Harris pointed out, 'Some station-hand may come along afterward and not being so squeamish obtain the scalp and tail.' As Harris described, doggers were permanently ingrained with the smell of death: 'A man's hands will smell for days after handling a deceased dingo – even if it has been dead only a matter of minutes – no matter how hard you scrub with soap, hot water and carbolic.'

Johns and his boys were essentially what were known throughout the bush as 'sundowners', itinerant bushmen who wandered through the outback, turning up unannounced at homesteads at the end of the day looking for a feed. Sometimes sundowners were well received, as they brought news of the outside world to far-flung outposts. But at other times, for families struggling to find food to put on the table, these sundowners were a downright pain.

Johns told Blakeley he had seen some of the drums of water the expeditioners had deposited along the route: 'I lost no time in getting down here, since we are practically out of tucker. We've been living on kangaroo.' Blakeley, the old overlanding cyclist, knew their predicament only too well – at times he had been grateful for any food or water anyone could spare. Nevertheless, every handout the expedition made lessened its self-reliance.

Johns had with him something the expedition didn't have, and – as it was turning out – they badly needed. Camels. So far, the greatest impediment to the expedition's progress had been its supposed great advantage, technology. They had lost so much time due to the frailties of the trucks and the worthlessness of the *Golden Quest* that Blakeley was despairing of find new ways to keep momentum. There was still no sign of Colson, who was bringing much-needed fuel and water in the Chev truck, and how much longer the battered and stripped-down Thornycroft would last was anyone's guess. They were about to try to travel across serious sandhill country, but in their intended direction they would no longer drive along the sand channels. Instead the truck would need to climb the dunes head on, one after another. Blakeley had been told by the old-timers in Alice Springs, who doubted the capabilities of the Thornycroft truck, that the camel – or Arabian dromedary as it was called – was the only way to traverse the desert. Before their departure, Blakeley had even requested in his report back to Sydney that they dispense with the idea of using aircraft, instead getting hold of some 'fast-riding camels', but his suggestion was rejected.

Precisely whose idea it was isn't known, but a deal was struck between Blakeley and Johns to supply him and his boys with tucker – 'whatever provisions he wanted, also tobacco' – on the proviso that they wait at Ilpbilla for a few weeks while the expedition attempted to get through to where Lasseter needed to be to take his bearings. Blakeley wrote that Johns agreed to 'stay around a while' and the pair drew up a chart with dates and times when the expedition would send up 20-minute flares should they require Johns and his camels: 'with these five camels on call it meant that if we got into country we could not get the truck through we could use the truck as a sub-base and get almost anywhere.' This was what it had come to. What Blakeley had once described as 'the best equipped turnout that had ever been in this country' had now degenerated into an ad hoc lame duck, horse trading with a dogger.

Johns's Aboriginal boys kept to themselves. One of them, Rolfe, spoke remarkably good English, something Blakeley commented about to Johns. Rolfe could also speak, read and write in fluent German, a skill he had learned at the Lutheran mission at Hermannsburg. He spoke several tribal dialects too, all of which impressed Blakeley mightily. As it transpired, Rolfe happened to be married to a woman who came from this region, and he knew something of the language spoken by the Aborigines who had wandered in from the desert. Blakeley asked Rolfe if he would try his hand at communicating with the light-skinned natives, and after considerable time spent conversing and drawing in the sand, Rolfe returned. 'These light-coloured people,' Rolfe told Blakeley, 'they belong to the sand-hills country.' But they would not tell him if there was any water in the direction they were headed. They did not want them to go that way.

Blakeley learned that Johns and Rolfe had spent time working for Bob Buck, clearing the scrub at Ilpbilla to make the aerodrome for the Mackay Aerial Survey Expedition. He asked Rolfe if he knew anything about Rip Van Winkle and the women and children in the other Aboriginal group, to which he replied that 'Rip' had fathered the children of both the women, who were his daughters. 'Rolfe was a very religious boy and thought it a great sin for Rip to go on like that.'

The expedition's crew was now mentally prepared to undertake their next long truck voyage to find a point determined by Lasseter to be 150 miles south of where he had changed his plans at Mount Marjorie. Yet they were not ready to depart. They were running low on both petrol and drinking water and there was still no sign of Colson and his Chev truck bringing their supplies. The feeling was that something must have happened to him on the drive out from Alice Springs. For the moment, Lasseter's journey south would just have to wait.

•

'It was not palatable to any of us, but in the circumstances there was nothing else to do,' wrote Blakeley of their decision to mount a search and rescue mission for Colson. They were postponing the big drive south to Lasseter's new location, and it was now the fourth time the Thornycroft had travelled the road they had made to Ilpbilla – returning east once again. 'We were all silent and felt how useless the trip was.' They had left Johns and his boys at the old aerodrome, where they would wait until the Thornycroft returned to resume the search for the reef.

Once more the truck was being inexorably drawn back to the hoodoo zone that was Aiai Creek. Lasseter, who had not seen the spot where the *Golden Quest* had crashed, wandered around the site trying to piece together what had happened while Taylor and Blakeley refilled the Thornycroft's fuming radiator.

The track was so much easier to negotiate than when they had carved through it the first time. It took three weeks to break through it then, but now they could be back to Alice Springs in two or three days. They made it as far back as the Dashwood by nightfall – and it was during the evening meal when they suddenly heard the drone of the Chev truck. Everyone looked at each other, relieved. 'I looked in the pot. Yes, there was still plenty there; we knew he would be as hungry as we were,' remembered Blakeley, thankful his mate was okay.

The expedition members were desperate to know what had happened to delay Colson for so long. The bushman wasn't quite his usual self and he had emerged from the truck with an apprehensive look, clutching a stack of mail. It was as if he was bringing bad news. For a moment it flashed across everyone's mind that perhaps it had something to do with Coote's recovery. Colson sat on the ground and began to eat his meal, explaining that when he had returned to the Alice with the wreckage of the *Golden Quest*, he had set aside a couple of days to do a few things for his wife before heading back out to meet the expedition. As he was just about ready to leave, he said, 'I received word from the Sydney crowd that they did not require my services any longer.'

Blakeley immediately smelled a rat, saying, 'Isn't Coote in the Alice?'

'Yes,' said Colson, 'but I think he knew before I did.' He then told of how he telegraphed the Sydney office, telling them that the expedition had now travelled so far to the west and they were anxiously relying on him to return with supplies. He said he also told Coote of this, but it was of no use.

'Finally the Sydney crowd closed the argument by telling me that I was sacked, and would not be required anymore. They told me they were running the expedition and would manage it as they saw fit.'

Blakeley could feel his blood pressure rising.

Colson said he went to see Ernest Allchurch, the special magistrate, who then went to the Government Resident, Vic Carrington, who advised them to wait and see if the Sydney office intended to send out another truck. Carrington asked Colson how long the expedition's supplies would last, to which he replied about three weeks, unless something went wrong. Colson warned them that 'if the truck caught on fire . . . it would depend on how long they could exist. I told them every precaution was being taken against fire but one never knew when a drum of petrol would spring a leak.'

Allchurch then confirmed that the Sydney office had no intention of sending another truck. Instead they were sending another plane. Colson couldn't believe it, and confronted Coote to tell him that 'I was going out on my own initiative and would make a fast run and tell the party what had happened, I informed the magistrate and he thought it was the only thing I could do.'

Blakeley was beside himself with rage, and rent the desert air with a tirade of expletives. The board in Sydney had sacked the most knowledgeable bushman on the whole expedition, cancelled their only support vehicle, and were bringing out another damn plane!

'After hearing that story,' Blakeley wrote, trying to downplay his outburst, 'I am afraid I said a lot of things; my blood fairly boiled for a while. All I said was quite in order and more than justified; for those boneheads to calmly and mercilessly condemn us was shocking.' He discussed his next course of action with Sutherland and decided to re-employ Colson despite whatever order had come from the Sydney office. He then sat down to write 'my report to the Sydney crowd; it was hot from the word go', he recalled. It must have been quite something, as Blakeley's vitriolic first attempt needed to be rewritten after he had calmed down, but the message wasn't lost.

He outlined the agreements he had made with the company and said that in future they were not to take notice of any information unless he had undersigned it, and for them to 'get it out of their heads that a plane could look after a big truck'. He wrote, 'The boot was very much on the other foot; in fact it would take two trucks to look after the plane.' He added that the decision to sack Colson could well have 'resulted in the whole party perishing; also that the hot weather was coming in on us, and that their interfering had cost them two weeks of good weather we could ill afford'.

The expedition's leader was well aware that his chief problem was the wily antagonist located in Alice Springs. He was certain that Coote had been intercepting messages, and knew he had unfettered access to Sydney by way of the telegraph office and the postal mail. Blakeley's only method of contact with the outside world was a makeshift Pony Express system using a Chev truck – thanks to the now sacked Colson – which had been able to ferry communications to the Alice and back. Blakeley gave Colson a letter for the postmaster instructing him not to give the mail to anyone but Colson, who was appointed as the leader's representative.

In the mail Colson brought out from Alice Springs was a note from Coote advising that another aircraft would arrive in several days. Blakeley shook his head. What would have happened to the aircraft and its occupants if the Thornycroft had successfully continued on west and hadn't turned back? No question, they would have perished.

Colson was ready to turn the Chev truck around for Alice Springs once more, and being so close to Archie Giles's station, Blakeley agreed for Mickey to leave with him. The Aboriginal guide had been away from his family for far too long, and ever since the expedition had entered country into which he had never been before, Mickey had been ready to go home. Despite Blakeley's earlier reservations concerning his eyesight, Mickey had proven as hard a worker as anyone in getting the vehicles through, and an invaluable extra pair of hands. They gave him a tuckerbag full of supplies, and Mickey wanted to sit with it in the back of the truck. 'Him tuckerbag might jumpem off,' were Mickey's farewell words as Colson drove off, Blakeley noted.

'Again the trucks parted,' recalled Blakeley, 'each going in opposite directions.'

•

Their arrival back at Ilpbilla brought with it a heavy air of frustration. Another bout of interminable waiting and counting the hours – this time for a second aeroplane – had raised Blakeley's ire to a new level. 'I was making vows, both aloud and silent, that once a plane came I was going to call the tune. It would have to run according to plan. I was going to end outside control.'

In the afternoon, Blakeley and Sutherland decided they would put the pressure on Lasseter to tell them what he really knew about the reef's location before the plane arrived. They asked him whether he remembered what the country was like where it was found, to which he replied that 'the walls were soft, slaty schist. He said it was only here and there it was outcropped,' wrote Blakeley. 'There were many shifting sandhills and odd patches of mulga scrub. He said he and Harding walked for five or six miles along it picking up stones at random. When the bag got too heavy they broke the stones up, then quartered them down to about a pound of stone. They did this all along the outcrop. This was the sample they took back and got assayed.'

What Blakeley was attempting to do was find out what Lasseter actually did in the Western Australian goldfields for the three years between when he first found the reef and when he returned with Harding. What sort of work did he do around the diggings? 'We never succeeded in getting any information other than it was any odd job; work was easy to get,' Blakeley wrote. It was hopelessly frustrating.

Lasseter was getting on Blakeley's nerves, adding to the torment of the ever-increasing heat and the bush flies attacking his eyes, ears and nostrils. The prospector had 'taken on one of his religious turns', as Blakeley put it, and began singing hymns non-stop. 'I like to hear people singing no matter how badly they sing,' Blakeley wrote, 'but when he has to listen to a score or more hymns all sung to the one tune, it's a bit trying. If I walked about in the sun the flies would nearly eat me. If I came back into the shed I had to listen to his moaning.'

In time, the four men begrudgingly resigned themselves to the wait. Occasionally Paul Johns and his boys would come and go – secured to remain in the vicinity by their loose arrangement with Blakeley – and the sandhill men with their two boys returned as well. Blakeley and Sutherland taught the kids how to play noughts and crosses in the sand. And, as Blakeley put it, 'The first and second days of waiting for that fool of a flying machine were not too bad.' In an attempt to relieve the boredom, Blakeley, Sutherland and Taylor began swapping stories, Sutherland asking Blakeley about his time on the opal fields out on the Stuart Range. Blakeley spoke of his tough and cruel experiences when mining out there in the old days. Of how he and his cycling companion Dick O'Neill had had their lease confiscated by the government and the three miners who then took it over struck it rich. 'We were merely the mugs who had done the hard work.' He spoke of infernal temperatures – of how he had been on a 125-mile-long bicycle ride and suffered appalling sunstroke from the heat reflecting from the ground. And of two old miners who perished out there: 'their bodies were never found . . . it was no use looking for them for the dingoes would have destroyed all traces. A pack of hungry dingoes can eat all the bones of a bullock, except the hard leg bones, and the strong jaws of the wild dog would easily crunch the bones of a mere man.'

But then, in an intriguing segue, Blakeley began to tell Taylor and Sutherland a story about a prospector, 'who came in with a few gold specimens'. As Blakeley described, 'He was a dying man, all dried up inside. Another fellow looked after him until he died, and the sick old man told this chap of the big reef he had found, so long that he could not see either end of it. He told a terrible tale of how his black boy had died of dysentery, how his pack camel had given in at last and he had to shoot it. He had to leave his bag of gold and managed to save only the few specimens. He gave these to the man who cared for him, together with a map of the reef, and then he died.

'For 30 years this other man talked about the dead prospector and the reef of gold until he was so bitten by the gold fever that he had come to believe the story and went searching for the reef. After a few weeks of wandering about, he too went completely mad and came to another man's place to die where he told the story all over again as if it had happened to him and he had found the reef. These poor fellows gather a few specimens from somewhere and they keep them for years, gradually building up a yarn that they consider themselves to be true.' Blakeley told Sutherland and Taylor that that was what he thought had happened to Lasseter: 'He had told his yarn so many times that he had come to believe it himself.' Whether Blakeley had genuinely heard this tale or whether he had invented it as an allegory about Lasseter isn't known. But what is certain is that fables about a hoard of gold lying somewhere in the middle of Australia waiting to be claimed had been whispered about since the first convicts were rowed ashore in Sydney Cove.

26

GOLDEN QUEST II

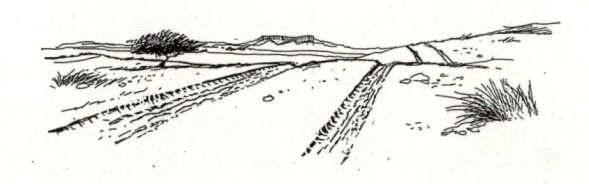

At some point on the flight out from Alice Springs to Ilpbilla, the DH60 Moth biplane Pat Hall had flown out from Sydney to replace the *Golden Quest* was imaginatively named the *Golden Quest II*. It was now buffeting through air pockets high above the McDonnell Ranges. Hall and Coote had only recently taken off from Alice Springs, but even before the flight commenced things had been problematic. Coote, still suffering from his injuries, needed to be driven across the aerodrome to a position alongside the aircraft, and physically lifted into the cockpit.

Hall had overseen the biplane's takeoff from the Alice Springs aerodrome and then handed the controls over to Coote. 'Once more I had the pleasure of handling a plane.' Coote figured that a lot must have happened since he had been taken to hospital – but in reality, for all the grinding around the desert in the Thornycroft, the expedition was no closer to finding the gold reef than they had been when they left Alice Springs. After all this time, Lasseter was still leading everyone around by the nose, but his cagey, erratic disclosure of snippets of information was starting to wear thin.

In time the little aircraft was approaching the Ehrenbergs, Hall frantically tapping the fuel gauge, aware that the tank overhead would be almost out. Coote wasn't perturbed: '[Hall] thought we must be getting low, but I told him that the needle in the gauge had only just disappeared so we had nearly another hour's flying available.' It seemed Coote's whole life was conducted along the same principle. Circling the Ehrenbergs, Coote was busily looking for the Ilpbilla landing strip. He knew it was somewhere on the north-eastern side of the range – but where? With fuel and time running out, Coote began to panic until he spotted the Thornycroft's tyre tracks in the sand far below. And there was the aerodrome, half hidden in a basin between two mountain spurs. The landing came in the nick of time – there was less than a gallon of fuel left.

Blakeley walked up to the taxiing aircraft while the switches were shut down and the propeller came to an eventual stop. He stood with his hands on his hips, disapprovingly looking over the fuselage from end to end, his expression announcing he was unimpressed with the replacement machine. 'When I examined the plane, comparing it with the pretty *Golden Quest*, it looked dirty, scaly and scabby; it had not had a coat of paint for years.' He was introduced to Pat Hall, the pilot, and was bewildered as to why he would risk his neck flying it. 'I immediately asked him where the hell he got such a bottle-oh-looking thing and he said it was all right, it got you there and back.'

Blakeley and Hall lifted Coote from the cockpit.

Hall, unaware of the tension the new aircraft had created between Blakeley and Coote, let it slip that he was glad they weren't travelling any further because they hardly had any fuel left. 'She only does about sixty miles per hour all out' – hardly what Blakeley wanted to hear.

'Well, to think those boneheads in Sydney sent a thing like that,' he scoffed. 'You cannot be overloaded with brains to venture out in a thing like that.'

Coote interrupted, telling Blakeley that they were going to fly back to Adelaide to fit the new place with the bigger engine from the wreck of the original *Golden Quest* and add an extra fuel tank, giving the aircraft much more speed and greater range.

Flying back to Adelaide to fit another engine and another fuel tank? Why the hell didn't they have all that sorted before they came out? Ilpbilla was not exactly en route. For Blakeley, the only good news in all this was that Coote was leaving and taking the plane with him.

Coote said they flew out because he thought they might be of some assistance, suggesting that Blakeley could 'look at the country you are trying to make for'.

'Well, you can get it out of your head if you think I am going to fly in that thing from the Ark,' Blakeley let them know. 'If you two are not endowed with any more sense than to risk your lives in it that does not say I have to act the goat too.'

They set about tying the aircraft down, an exercise the frail Coote had no chance of helping with. It was obvious that he was still in a bad way from his accident and clearly should not have attempted the flight out. His injuries made him somewhat tetchy and his eyes locked onto the two boys from the mission.

'Where did the Aboriginals come from?' he asked.

'They belong to a fellow from Hermannsburg Mission, Paul Johns,' replied Blakeley, and told Coote how lucky Johns was to find them. 'He has been out dogging and, all in, just made our camp.'

Coote asked Blakeley whether the tall young man in the pale green shirt walking towards them was Johns, to which he said yes. Coote didn't like it. His record of his conversation with Blakeley about seeing Johns for the first time is particularly interesting.

'What's he doing here looking for dogs?' Coote asked. 'The last dogs we heard were back at Haasts Bluff. Have you heard any round here?'

'No, can't say that I have. But what harm is he doing here?'

'Nothing at present,' Coote replied, 'but this is an aerodrome, not a wild-dog trapping place. Let him get out. I'll bet he's hanging around to either join or follow the expedition. Did you say he came from Hermannsburg Mission?'

'Yes,' said Blakeley.

'Well, that is only about seven days' camel ride from here so he ought not to have run short of food so soon. Anyway, if he was getting short why did he come here? Why didn't he turn back to Hermannsburg? He knows there's no store here.'

Blakeley agreed, saying he had noticed that when Johns turned up he seemed in a bad way, but now, all of a sudden, he had recovered and looked 'in good nick'.

'I don't care what he looked like,' said Coote, 'but somebody's going to be sorry if he isn't out of here by tomorrow, or before we leave.' Coote made his opinion of Johns clear to Blakeley. 'I'm suspicious of that guy. He has improved too much. Give him some good advice. Tell him there is some good trapping in the Haasts Bluff country. That's good country; this looks like the world's worst to me.'

The mail Coote had brought with him from Alice Springs did nothing to improve the situation. Word had come back that Captain Blakiston-Houston had decried the behaviour of the expedition members, saying they were 'quarrelling like a lot of schoolchildren'. The new instructions from Sydney to Blakeley were even more startling. From these, everything seemed to involve Blakeley supporting Coote and the aircraft: 'the pilot would do this, the pilot would do that,' he recalled.

Anxious about what Blakeley's response would be to this aviation-focused missive, Coote asked him if he had read the instructions, to which he angrily replied, 'Yes, but hold hard. All this looks like instructions from you. I told the boneheads in Sydney not on any account to send another plane, but to let you get some good riding camels and send them out. By these instructions it appears that you have been appealing on my behalf to hurry up and send another plane.'

That Coote had been meddling with Blakeley's requests while he was in Alice Springs was in no doubt, and Blakeley was ready for a showdown. He pointed to the plane sitting in the red dust. 'If ever you come out here in that, bring your own organisation to look after you, for I consider you want two trucks and about four men to attend to you. You had a hide telling the Sydney crowd you could tender the big truck with a thing like that.'

It was clear to everyone that the expedition was now being pulled in four directions: Blakeley, who was struggling to make decisions on the ground; the impatient and unsympathetic Sydney office, acting on behalf of the shareholders; the Machiavellian Coote, who was deliberately manipulating and obfuscating communications between the two; and of course the volatile, secretive and paranoid Lasseter, a perennial conundrum who ultimately held the key to whether the mission would succeed or fail.

It was Lasseter who was keen to observe the lay of the land from the air. Taylor had checked over the rather shabby *Golden Quest II* and was satisfied of its airworthiness, and the following morning Hall and his passenger took off, headed south-west. Earlier, Blakeley had made an agreement with the pilot that the aircraft would follow a precise bearing to a precise distance out and back – no deviations. They had already had one aviation catastrophe – should there be another, Blakeley wanted to know where to find them. All they could do now was wait.

Always in the background was the German dogger, who somehow did not fit the scene. Perhaps it was his youth or perhaps it was his well-educated European air, but there was something about Johns that made everyone, it seemed, feel uneasy. Coote recalled meeting Rolfe, one of Johns's native boys: 'Two of the aboriginals were from the mission and had accompanied Johns on his dogging trip. One of them was a particularly fine type of Arunta, with the wavy hair and beard that is so typical of that tribe. He was well educated for an Aboriginal and he impressed me with his splendid air of independence. It was evident he was not very well disposed towards Johns.'

•

It was a long and nerve-racking two hours before the hum of the biplane's engine was heard again. The pair had flown as far as Docker Creek and then across the Western Australian border before turning around and heading back to the airstrip. Everyone was at Lasseter, bombarding him with questions. What did he see? Did he recognise any landmarks? Could the truck get through? How far did they fly?

To everyone's disappointment Lasseter confirmed they were indeed 150 miles too far north, but said he had found a way through the sandhills, drawing Blakeley a sketch of what he had seen from the air. But just how navigable was the country for the Thornycroft? Blakeley sent Taylor to fly with Hall for a second reconnaissance; they returned half an hour later and the young mechanic announced that he was not hopeful of getting the truck through but he would give it a try.

Coote had detected a distinct change in the expedition members' attitude since he had been gone – the raw enthusiasm they had started out with had leached away. 'Judging from the talk, the rest of the party, excluding Lasseter, was getting fed up,' he wrote. 'The heat was more intense, the truck was now more finicky, and enthusiasm for the task was rapidly waning. Lasseter's announcement that the party was such a distance too far north had been a staggering blow, so that afternoon I led Lasseter aside.' The pair wandered away from the camp – out of earshot of everyone – where Coote asked Lasseter if he had seen his landmarks from the air. His response was startling.

'Yes, I did,' he said, 'and what's more I saw the reef. It's there as plain as a pikestaff. We flew at only about thirty feet when we were near it. It is in the heart of the mulga and timber country. It was impossible to land there.'

Coote was incredulous, asking him if this was 'absolute gospel', playing on Lasseter's Mormon proclivities.

'Absolutely,' he replied.

Coote was dumbfounded. Lasseter had *seen* the reef. He actually saw it – he flew over it! But why didn't he say anything – why did he just pretend that he had seen nothing? If he had recognised landmarks, he was contractually bound to notify the expedition. When reminded of this rather important detail, Lasseter complained that he had no confidence in Blakeley, was wary of Sutherland, and particularly did not trust Colson one iota. Coote said he told Lasseter that his 'likes and dislikes did not enter the matter; you have an obligation to the shareholders'. He continued, saying that if Lasseter didn't reveal his landmarks, they would fly him back to Alice Springs, where 'he would probably face arrest for bringing us out here on a wild goose chase'.

Lasseter acquiesced; sitting on the ground he picked up a stick and drew a diagram in the sand. He pinpointed the position of Lake Christopher across the Western Australian border, describing it as a radial point. He then drew three hills, which he said were unmistakable – referring to them as 'The Three Sisters' – 'like three women in sun-bonnets talking to one another'. About another 35 miles south-east was another hill, 'shaped like a Quaker hat – tall, conical in shape with the top cut off'. He said the reef lay about 10 miles east from a lakelet, and if you looked along the reef in a north-westerly direction you could see that these Three Sisters appeared to line up with its far end. He believed a plane could probably land on the lakelet but the reef country itself was heavily timbered. Yet the reef could unquestionably be seen from the air, 'just peeping through the mulga'.

'There you are,' concluded Lasseter. 'Are you satisfied now?'

Coote remarked that at least that was something to go on – yet he was sceptical, recalling that while trawling through copious maps back in the Sydney office, Lasseter had emphatically denied that the lake supposedly near his reef was Lake Christopher. Then again, wrote Coote, 'That may have been done only to put us off the scent.' With Lasseter, who knew where his paranoia would lead you.

The prospector begged Coote not to tell a soul what he had just revealed, saying that he would tell Blakeley in good time.

The pair walked back to the campsite where Blakeley, suspecting something underhanded had been going on, called out to Coote, 'Been getting the good oil?'

'Just talking aeronautics,' Coote replied.

So now two people on the expedition knew the whereabouts of Lasseter's reef – but not the expedition leader.

•

It still struck Coote as odd that Lasseter was keeping quiet about his rediscovery and so he questioned the only other person aboard that flight south-west – Pat Hall, the pilot. 'He claims he found the reef,' said Coote. 'What do you think about it?' Hall wasn't sure, but he recounted how at about an hour and ten minutes into the flight Lasseter 'began to jump about in the front cockpit very excitedly'. Hall continued, much to Coote's interest, 'He pointed at something but I could not hear what he was trying to tell me. He almost hopped out of the plane – there was no doubt he was genuinely excited. Then he waved me to return.'

Coote was suspicious. Lasseter had told him he had actually seen the reef, but he had told everyone else they were 150 miles too far north. The flight out and back had lasted little more than two hours, but there was no way this Gipsy Moth could travel at 150 miles per hour. In fact, its cruising speed was officially about 72 miles per hour – and as Hall had said about the well-worn *Golden Quest II*, 'She only does sixty miles per hour all out.'

Coote asked Hall whether he had seen a small lakelet, as described by Lasseter, to which the pilot said yes. He thought you could probably even land on it. It was all too annoyingly inconsistent for Coote – maybe they had flown out 80 miles or so and Lasseter saw his Three Sisters in the distance, 'one of his landmarks that was as good as seeing his reef'.

The following morning Coote and Hall were to return to Alice Springs and then fly the *Golden Quest II* to Adelaide for the engine refit. Lasseter wanted to send a message back to be telegrammed to his wife and approached Blakeley to receive the okay. As per the arrangement, Blakeley read it and gave it his approval. The pilots were helmeted and goggled and positioned in the cockpit when Lasseter walked over and handed it to Coote. It would be the last time they would see each other. Taylor gave the propeller a few swings and Hall soon called that the ignition had contact, the engine fired into life and the plane sped along the red earth, its tail waggling as it climbed into the air. 'I felt jolly glad to see the end of Coote,' wrote Blakeley.

The truck was underway almost as soon as the plane had left, but the going was dreadful. Out came the old coconut matting to place under the wheels. The mats were now in a very poor state, hardly helping with the big truck's progress. But at least Blakeley had the sketch Lasseter had drawn up after his flight to go by. He had said that from the air the going looked pretty good, as it was mostly saltbush, and from their experiences that kind of country was ideal for travelling. 'But Harry's saltbush was impassable,' said Blakeley, 'so we still had to carry on walking ahead.' The next few days were much the same, the going painfully slow. Short, hard sticks in the spinifex frequently punctured the big tyres. Blakeley would regularly stop to take compass bearings, and every time he did so he included Lasseter in the process. 'I got Harry to give me a hand with this because I did not want any arguments about our position when we got 150 miles south of Mount Marjorie.' He was leaving nothing to chance.

It was then that Blakeley fell ill. Sweat poured down his face and he trembled so much he could hardly stand. Everyone agreed he looked in a bad way. The following day Sutherland suffered the same complaint. 'We were not having sufficient food,' wrote Blakeley, and a decision was made to stop earlier and cook proper meals. 'We were always too tired at the end of the day to cook, so just had any old thing, mostly out of tins.'

He wrote that by this stage they had lost count of the days. Lasseter's rough sketch was more or less right, and every now and then the prospector would tell them to stop the truck and he would set off to explore an area on foot. Blakeley would go with him, but nothing ever eventuated from Lasseter's wanderings.

One night Blakeley was drinking tea by the campfire when Sutherland walked over to tell him that Lasseter was lying down. 'What's the idea of the big walk today?'

'I'm puzzled myself,' said Blakeley. 'I went where he wanted to go, God only knows why he wanted to climb over a dozen sandhills.'

Lasseter ate nothing that evening. When asked if he was okay, he said he was too tired to eat. It was the first night on the expedition that he didn't write in his diary. 'He was very quiet,' Blakeley noted. 'Before this if he were quiet, it was a hundred to one that he was sulking and would snap at anyone over nothing; but now he seemed pleased we were taking notice of him.'

27

LOOKING FOR THE DOGGER

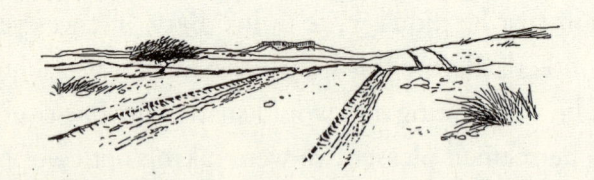

That night, they worked out they were almost 20 miles from the Western Australian border and close enough to be 130 miles directly south of Mount Marjorie. According to Blakeley's calculations, they would very soon be standing on the spot where Lasseter said he needed to be.

The following morning they set off early to press on that much further, but the going was so rough that by the afternoon they had covered only 10 miles, and they decided instead to set up camp for a few days. They had noticed that their path had been gradually rising, and a distance from where they were camped they could see a vast sand dune, its top flattened by the desert wind. Blakeley knew that this would be a perfect vantage point from which they could pick the route forward, but the four men were just too tired to slog their way to the top that day.

Blakeley woke early and began the long walk to the sand dune's summit, only to be confronted by an overwhelmingly demoralising view. Before him the ground dropped 300 feet into a vast, heartbreaking depression of busted, twisted terrain that stretched for miles, impossible for anything, let alone the Thornycroft, to cross. Blakeley described it as 'the strangest sight of tumbled, tangled country. I knew when I looked over that we had come to the big breakaway, and knew that the truck would never look over into that valley.' If somehow the big truck did crash and stumble its way to the valley floor below, there was no chance of it ever climbing out. The rest of the expedition scaled the dune to see for themselves, all flabbergasted at what stretched out before them. Sutherland remarked, 'We can get down there but she will never get out.'

In the distance was the faint outline of the Petermann Ranges, their access cut off by this impassable geological fissure. For the expedition, it might as well have been the Grand Canyon.

This was the big breakaway country Blakeley had warned Lasseter about, knowing full well that if he had ever set foot out here and actually seen it he would never have forgotten it. Blakeley explained to the others that, as far as he knew, the continent of Australia began as a kind of flat-topped mountain protruding from the depths of the ocean floor. Its edges were giant cliffs. Millions of years ago, the part they now knew as the Great Australian Bight had slipped into the sea, causing vast landslides inland, such as the 'big breakaway' they were looking at now.

Blakeley had been waiting for this moment since Lasseter took the sextant bearing from the top of Mount Marjorie all those miles ago. Since then he had followed the prospector's wishes to the letter, taking him wherever he wanted to go. Now the journey had come to an abrupt, unmistakable – unquestionably final – dead end. Blakeley asked Lasseter if he'd like to take another bearing with the sextant – but as expected, he declined. Lasseter was snookered.

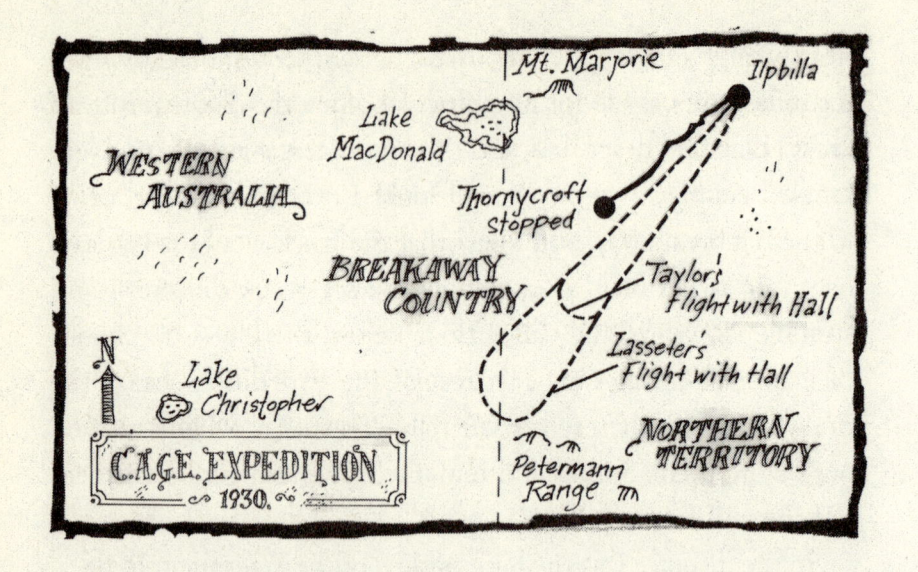

[Map: C.A.G.E. Expedition 1930 — showing Western Australia, Lake MacDonald, Mt. Marjorie, Ilpbilla, Thornycroft stopped, Breakaway Country, Taylor's Flight with Hall, Lasseter's Flight with Hall, Lake Christopher, Petermann Range, Northern Territory]

'Well old boy, are you satisfied now that you're bushed?' asked Blakeley, triumphantly, as though he had just cracked some Scotland Yard case. He was confident that very soon he would be returning to Sydney with the indisputable evidence that Lasseter had never ever been out here before. 'No animal, horse or camel could cross that country.' Like the closing chapter of a Sherlock Holmes novella, Blakeley began rattling off all the inconsistencies in Lasseter's well-worn tale. He pointed out that Harding's bearings taken all those years ago could not be south where Lasseter indicated, as that would make the total distance 1300 miles. 'In all your stories,' said Blakeley, 'you have never been more than six to seven hundred miles.'

Lasseter was flustered, his speech staccato. 'I don't know what to think about the distance, but my instinct is that it's over there.' He pointed to the violet haze of the Petermann Ranges, Blakeley recalled. But even though the Petermanns were within sight, to get there meant a phenomenal amount of backtracking, a round trip out there adding another 500 miles.

Blakeley was convinced that this was just another ad hoc Lasseter ruse – an impulsive directive thrown up when his last dubious instruction fell apart. It seemed that Lasseter was now throwing out ideas as to where his reef was as if Central Australia was some huge dartboard. Blakeley had heard the old 'prospector finding gold in the desert' stories over and over since he was a boy, and he was appalled that this one had actually taken him in. He knew of many expeditions that had gone out looking for mythical reefs just like Lasseter's. Blakeley expressed his disappointment: 'The reason why I listened to your tale in Sydney was because it was different to all the tales I had heard before,' he told him.

Lasseter admitted that he had never seen the breakaway country before. If he and Harding had passed through here, there was no avoiding it. This was as close as Blakeley came to obtaining a confession from Lasseter that he had made the whole thing up. The expedition leader drove home the point of just how ludicrous Lasseter's claims were, especially concerning the incredible variations in the distances purportedly travelled to and from Carnarvon. 'Anyway,' said Blakeley, satisfied that Lasseter knew the jig was up, 'why select Carnarvon? There are lots of places closer than that.'

Lasseter didn't respond except to ask what Blakeley intended to do next. He replied that with the hot weather now upon them, it was too late in the year to do anything but return to Sydney and present his report to the board. Incredibly, it had come to this – Blakeley was finally calling the expedition quits.

•

Whether Lasseter's reef truly existed somewhere out there was now immaterial. The expedition would soon be over, and Blakeley was heading back to face the Sydney crowd empty-handed. But what would Lasseter face on his return to Sydney? Ridicule, ostracism – his family shamed most likely. But it was the unforgiving wrath of John Bailey that did not bear thinking about. The most powerful union boss in the country had poured thousands of pounds of union members' funds into this expedition on the basis of Lasseter's flaky story, and when the day of reckoning came he would want his pound of flesh. If the shareholder-subscribed company dissolved into bankruptcy, you could guarantee it would not be the company chairman 'Ballot Box' Bailey who would be spending time in debtors' prison behind the stone walls of Long Bay Gaol. Lasseter would be strung up and hung out to dry – very likely charged with fraud and convicted.

Lasseter's time as a youth in the reformatory gave him firsthand experience of what life behind bars was like. And if it wasn't prison it'd be certain Bailey knew of blokes outside who would make sure Lasseter didn't pull a stunt like this ever again. Blakeley understood the dire consequences for Lasseter should he return. He had an idea, albeit a risky one – a possible way out for the little man who had painted himself into a corner. He was giving Lasseter the opportunity to do the 'honourable' thing. 'You have one slender chance,' he advised Lasseter sternly. 'If we can get this German dogger in, you will be able to disappear and save yourself and your family from disgrace.'

•

From Blakeley's point of view, his suggestion for Lasseter to 'disappear' would not have seemed too far-fetched a proposition. In his time on the opal fields, Blakeley would have no doubt seen all manner of questionable, itinerant men come and go from the diggings over the years, and would have shared more than one shot of liquor with blokes who had perhaps 'disappeared' several times in their lifetimes. It was no secret that some of these reclusive hard-bitten miners were running away from other lives, on the run from whatever or whoever was hounding them. Alimony, child support, bankruptcy, fraud, absconding, military desertion, armed robbery, even murder, would often see some new-chum or misfit arriving to try to carve out an anonymous new life on the opal fields. And no questions were ever asked.

As easy as it was for someone new to appear, if they stepped out of line it was even easier for them to disappear, the opal fields littered with mineshafts, working or abandoned, in which a dumped body would never be found. Far-flung mining communities were sometimes lawless, often considered places to avoid. Even during World War II when Australian and US army truck convoys were mobilised from Alice Springs and Mount Isa to supply the besieged frontier post of Darwin, service personnel were not permitted to stop at Tennant Creek; so violent and unpredictable was the reputation of the Territory's wild gold miners that the town was declared 'off-limits'.

In an age long before computerised records, it was relatively easy to 'disappear', and in the financial wreckage created by the Great Depression, it happened more frequently than anyone would dare admit. As it was, Lasseter's whole life seemed to have been a rolling act of disappearing and reappearing. When pushed, he had even admitted to using fake names, and Blakeley himself had seen his American passport using an alias.

That evening, it was decided to hold a sort of kangaroo court between the four men, whereby each had an uninterrupted say about what had transpired on the expedition. It was in effect a chance to air their grievances, in particular about Lasseter. Blakeley recounted all the inconsistencies he could think of: the ever-changing line-up of characters who had apparently rescued Lasseter (Afghan cameleers, Harding the surveyor, a dingo scalper); his lack of recognition of any landmarks; the story of his bearing reading on Mount Marjorie being the same as those placed in the bank vault in Sydney; and the inexplicable changing of directions and distances. 'Now you want to go another two hundred miles further away, making the total distance 1250 miles,' Blakeley complained. 'Worst of all, your only excuse now to go on is that you have a presentiment that it is over in the Winter Glen country; and that is the reason I intend to end the expedition.' Blakeley later remarked on this: 'I've always found that when a person is governed by presentiments or instincts they are grasping at straws for it is no better than taking notice of a fortune teller.'

'I then invited Harry to put up a case why we should go on but he had nothing to say,' recalled Blakeley. The expedition had ended. As for Lasseter, he still had one last chance, Blakeley informing him, 'I will report back to Sydney. If we can engage the dogger, I am willing to let you get away with him, then it will be up to the Sydney crowd to recall you.'

•

With the bough sheds of Ilpbilla now in sight, the remnants of the expedition were surprised to see a large mob of camels in the shimmering haze and a group of people walking about the airfield. They turned out to be Aboriginal employees of Bob Buck. The leader was 'a flash fellow', as Blakeley described him, named Billy Buttons. These boys had been sent out with a string of 30 camels to pick up the empty fuel drums discarded by the Mackay Aerial Survey Expedition – Buck thought it was time to collect his property before it was picked up by one of the expeditions wandering through here.

Blakeley eyed off the impressive camel train. What were the chances of finding such a godsend right here at Ilpbilla, just when the truck was giving out? That evening Blakeley offered to hire 10 of the camels and an Aboriginal hand, but Buttons declined, saying he was already overdue, 'and I know I will cop it when I get back'.

In the course of the discussion, an intriguing exchange took place between Blakeley and Buttons about Paul Johns. The young Aboriginal man either knew him or at least knew of him, from when Bob Buck had employed Johns in clearing the airstrip at Ilpbilla. Blakeley wrote of the conversation: 'I asked if he had seen the young German dogger about anywhere and he said, "No, has he been here?" And when I told him he said, "How long ago?" I told him, and added that I was going to signal him in. I said that he might be in any day and this seemed to disturb Buttons; it was quite obvious that he did not want to meet him.'

Precisely why Buttons was disturbed about meeting Johns is not known. However, various accounts suggest that Aborigines within the area were wary of him. Missionaries at the Hermannsburg Mission had caught the 22-year-old German interfering with Aboriginal girls and had asked him to leave.

Certainly, in the Central Australian winter of 1930, Aborigines in this region especially were distrustful of armed white men wandering through their land. Hardly two years before the CAGE expedition set out, a 67-year-old dogger named Fred Brooks had been poisoning and scalping on a property called Coniston Station, directly north of Haasts Bluff. Due to the devastating and seemingly endless drought, tensions between whites and Aborigines had become greatly inflamed. Overgrazing cattle had stripped the land of vegetation and destroyed traditional waterholes. Aborigines were perishing from hunger and thirst; tribes wandered properties in the hope of finding water.

Around 7 August 1928, Fred Brooks camped near a tribe, and an Aborigine named Bullfrog had urged his wife to ask the old trapper for food and tobacco. Brooks agreed, but on the condition that she wash his clothes. When she failed to return that night, Bullfrog headed to Brooks's camp the next morning only to discover Brooks in bed with her, and in a blind rage attacked the trapper with a boomerang, severing an artery in his throat. Bullfrog and two others then beat him to death before attempting to hide the body.

Upon the corpse's discovery, Mounted Constable William Murray, the officer in charge at Barrow Creek, who also bore the title 'Protector of Aborigines', formed a mounted posse and undertook a punitive mission of indiscriminate reprisal that lasted several weeks, hunting down and killing Aboriginal men, women and children. Precisely how many people were murdered is unknown. At an inquiry into the massacre, Murray freely admitted to shooting 17. He explained that he killed for practical reasons.

The judge inquired, 'Was it really necessary to shoot to kill in every case? Could you not have occasionally shot to wound?'

'No, Your Honour,' Murray replied. 'What is the use of a wounded blackfellow hundreds of miles from civilisation?'

Today the number of Aborigines murdered by Murray's posse is generally believed to be somewhere between 60 and 110, but possibly much higher. At the time, a Northern Territory policeman said Murray boasted that the number of deaths was 'more like seventy than seventeen'.

The board of inquiry found that 'the shooting was justified, and that the natives killed . . . were on a marauding expedition, with the avowed object of wiping out the white settlers . . .' Constable Murray was feted by the community and the press, the *Adelaide Register News* describing him as 'the hero of Central Australia'.

Word of the massacre spread through the Aboriginal population, creating fear of white men that would last for generations.

•

Paul Johns's unannounced arrival the following morning was particularly disturbing; Blakeley wrote, 'His entry to our camp, in response to my signals, rather antagonised us all.'

At dawn, Johns had tethered his camels and left his boys half an hour away from the airfield at Ilpbilla and made his way there alone and on foot. While the expedition members were still asleep, he crept up to the site and into the camp, where he produced a revolver and shouted for everyone to put up their hands; Blakeley recalled him 'pointing it particularly at me'.

Still in his swag, Blakeley did his block: 'I gave him a piece of my mind that was not music to his ears. He said he had seen my signal but I think he was lying.' It had been more than two days since Blakeley fired his last flare, and had Johns seen it, he could have arrived at Ilpbilla within hours. The dogger's actions left Blakeley feeling uneasy about his plan. 'Had there been another way he would not have got the job,' Blakeley wrote.

Johns's inexplicable behaviour had set an uncomfortable tone for the day. But even without his questionable prank there was a palpable air of tension. For the expedition members, the thought that Lasseter could soon be leaving was in equal parts disappointing, worrying and a relief.

It was explained to Johns that the CAGE expedition was officially disbanding, but that the plan was for Lasseter to continue the search for his reef, and that he and his native boys and camels would be engaged by the company to guide him. They would head out, find the reef, peg it, and return.

A contract was drawn up, in which on behalf of the company Blakeley signed on the German dogger for two months, employed under the direction of Harold Lasseter for the sum of five pounds per week. If he required an extension, he would need to return to Alice Springs, where he would find instructions waiting for him. Blakeley also told him how he would be paid by the company: on demand, at the Government Resident's office. The agreement was signed by Blakeley and Johns and witnessed by Lasseter and Sutherland, dated 13 September 1930. Blakeley's signature was merely a formality. He no longer believed in the reef and was washing his hands of the whole distasteful show. As far as he was concerned, Lasseter was now free to continue traipsing around the desert making things up and changing his mind as he went.

In undertaking the camel journey, Lasseter could only take with him a minimal amount of personal belongings, and as expected, once again he began rummaging through his tin trunk. Blakeley glimpsed this from the corner of his eye. He had been waiting precisely for this moment, and he strode over and snatched the navigational almanac that had been loaned to Captain Blakiston-Houston. Lasseter was shocked, and pleaded with Blakeley to give it back, saying he could not continue on without it. Blakeley reminded him that he had promised to return it to its rightful owner, Rear Admiral Evans, and anyway, since he was not taking his sextant with him he had no use for the book. Lasseter was beside himself. 'He was dead nuts on taking this book,' Blakeley recalled, that night using it as a pillow.

'Next morning a hitch occurred,' wrote Blakeley. Johns's two Aboriginal mission boys had left during the night without food or water to return to Hermannsburg, 150 miles away on foot. Johns explained to Blakeley that they had refused to go with him. He said Rolfe had complained that four men with five camels would mean two men walking all the time – and they knew who those two men would be. Blakeley admitted, 'I did not realise they were going until Johns informed me they had gone.'

Now the party to find the reef comprised only Johns and Lasseter. A twinge of guilt flashed through Blakeley's mind as to the pair's chances of survival – the cocky young foreigner, and the eccentric old prospector who had never been able to prove that he had ever set foot in the outback before. And now they had no native escort. 'Most of the chaps who are good bushmen are useless without their native guides,' Blakeley commented.

Blakeley took Johns aside, warning him not to take any chances, and not to wander for more than three days from known water. Lasseter was still angry with Blakeley, once again demanding the return of the book, saying he had been 'tricked', because had he known that this book was not to go, he would have brought his own. 'I did not reply to this,' explained Blakeley, 'because we had not known that Captain Houston was bringing the book.'

It was just another pointless, circular argument between the pair. Lasseter complained that he had more right to the almanac than Blakeley, to which the expedition leader replied that if he was not taking his brass sextant, what good was the book to him? Lasseter insisted his Diploma of Survey was sufficient authority for him to continue with the almanac in his possession, but Blakeley held firm. Besides, the book was just added weight to what the camels were already lumbered with. Lasseter had already forgone his big tin trunk, now sitting on the back of the Thornycroft. Instead he had a smaller steel dispatch box, inside of which was a smaller case, both full of papers. According to Blakeley there was no room for the morocco-bound 'big fat diary' Lasseter had been writing in every night since the journey began, so 'it was packed away in one of the cases'.

Blakeley remembered the precise moment of Johns and Lasseter's departure: 'I packed all his personal effects on his riding camel; then gave him my own prismatic compass, since the one he had was very unreliable. I also gave him my company's watch and revolver so everything was all set. I shook hands with him and wished him luck.' The camels were swayed to their feet, braying raucously, their backs heavy with the two men's gear.

Lasseter turned to Blakeley, commenting, 'I will be able to prove to you later on that I was not telling you any dream yarn.'

Blakeley responded disparagingly, 'If you can find a reef over in that country you are a better man than I think you are, for of the twenty-two expeditions I know of, every one was equipped with the best prospectors in the land, and they found nothing.'

The camel train ambled out of the camp, the grand search for Lasseter's fabled reef now pared down to two lonely, disparate figures: the story's proponent himself and a young German dingo scalper for hire. They posed momentarily for one last photograph, with Lasseter standing ahead of the line of camels, the punishing sunlight creating a black shadow beneath the brim of his hat, obliterating his facial features.

Johns led the way out on foot. Harold Bell Lasseter was last seen walking wide from the last camel, eventually disappearing into the haze.

'We watched them out of sight,' remembered Blakeley, 'then I said, "That's the end of my millions," for had the ten mile reef been found my cut would have been worth about seventeen million pounds.'

28

ROCKS AND HARD PLACES

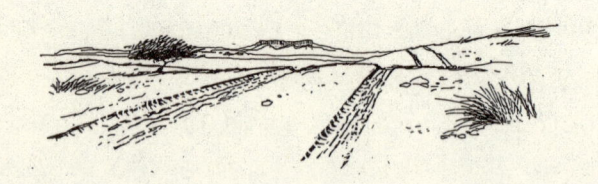

Blakeley had washed his hands of the expedition, and of Lasseter, and having packed up the Thornycroft, the remnants of the disbanded expedition spent the next few days preparing for the final drive back to Alice Springs. One evening they heard the faint, familiar whine of Colson's Chev truck approaching, and within 20 minutes he had pulled up. As he handed out the mail, Colson said he had camped with Lasseter and Johns along the track the night before. The pair had covered 30 miles that day. Blakeley recalled how everyone spent 'a quiet half hour' while they read their mail. There were the usual annoying missives from head office, but then, Blakeley said, 'In another letter I got a shock: Lasseter had wired the company that he had seen his country and located his reef. That was on the day out with Hall. This was the first I knew about it. I read that part aloud to the others and, like myself, they were dumbfounded.' Colson was amazed no-one on the expedition knew that Lasseter had found his gold reef, when it was common knowledge in Alice Springs. It seemed Lasseter had double-crossed Blakeley. 'The little religious blighter had not told me a word,' Blakeley fumed. 'He had tricked me with the wire to his family. I OK'd it but when he handed it to the airmen he could have palmed another wire.'

Compounding Lasseter's duplicity was the fact that the Sydney office had been deliberately bypassing Blakeley and the other expedition members on the ground. 'One would have thought that such an important statement from Lasseter would have brought more inquiries, especially with a flying machine on the job, that had, at least, got out there and back. They could have got a report from the two practical miners, Sutherland and myself, but they did not ask for that; they kept the information for themselves.' Blakeley was onto something. He smelled share manipulation. 'I did try to find out if there had been any dealings in shares but could not get the slightest information.'

As far as Blakeley was concerned, his instincts seemed to have been vindicated. 'Now I had the evidence I wanted, for here appeared to be a clear case of fraud,' he wrote. 'You can understand now why I had been so particular about anything that went over the air. I suspected that a stunt might be worked, but I was determined that I was not going to be a scapegoat. This telegram was clear proof that Lasseter intended to work a stunt on his own.'

In packing up they left a ton and a half of supplies at Ilpbilla, and put up notices 'telling white men what the food dump was in aid of'. They then set off from the makeshift aerodrome for the last time.

Colson drove ahead in the Chev truck. He had an appointment to keep back in town, but an arrangement was made that if the Thornycroft failed to return within a few days, Colson would conclude that the truck had broken down and he would drive back out to meet them. And break down the truck did. The big-end bearings were now starting to knock badly, signalling that the engine was now in its death throes. Taylor inspected the truck's filthy oil-spattered motor, and concluded sand had blocked the oil passages, starving the engine of lubrication. The machine had no air filter on the carburettor, meaning it was sucking desert sand straight into its internals. With a bit of stiff wire they had kept from the wreck of the *Golden Quest*, the pair took it in turns to twist it into the oil canals to clear the dust residue that had set like cement. It took two days to clear the blockages, Blakeley describing it as 'a rotten job'. But the noise of the bearings disintegrating only grew louder the further they drove.

'I can tell you it was dreadful to sit behind that engine when it was knocking so badly,' Blakeley remembered.

•

Errol Coote had flown the newly re-engined *Golden Quest II* from Adelaide as far as Oodnadatta, where the few corrugated-iron buildings that marked out the town were under siege from a violent sandstorm. 'That gale blew for three days bringing in its train huge clouds of dust that obliterated the landscape and filled the air to a height of 2,000 feet.' With the aircraft firmly tied down, all Coote could do was try to telegraph Alice Springs and Sydney that he was delayed. He received a reply from Colson in Alice Springs which he described as 'a veritable bombshell'. 'Expedition returning,' Colson wrote. 'Suggest you proceed direct to Sydney to avoid further expense.'

Coote then received a telegram from Ern Bailey in Sydney telling him that the expedition was heading back: Blakeley had given up but Lasseter was still out looking for the reef. The Sydney office now wanted to put Coote in command. They wanted him to go and find Lasseter.

'I felt savage,' wrote Coote of Blakeley returning without Lasseter. 'What was the idea of abandoning the man on whose knowledge the expedition had been formed? Why had the party not brought him back with them?' Coote wired Bailey back to say he was flying to Alice Springs as soon as he could.

Once in the Alice, Coote received the bare minimum of information about what had happened with Lasseter from a reticent Fred Colson; all he would say was that the prospector and Johns were now heading south to the Winter Glen area, out towards the Petermann Ranges. Coote, as the newly appointed expedition leader, needed to intercept them, and he had just the idea. He wired Bailey of his plan to scrap Ilpbilla as the base – they had only used it in the first place because of Blakeley's say-so. He would use Ayers Rock to the south instead. It made perfect sense, as there was reportedly a good water supply at the rock – the Mackay expedition said it was the best in Central Australia, and being 140 miles south-west of Alice Springs it was precisely at the gates of where Lasseter needed to go. But first Coote needed to do a bit of housekeeping to make sure everyone understood who was running the show.

In time the Thornycroft came within striking distance of Alice Springs. As Blakeley wrote, 'We saw the plane, evidently looking for us, for as soon as the pilot did a few fool things like dipping overhead, he flew back.' But the pilot, Coote, wasn't simply looking for them to check on their safety. He had another reason to find the Thornycroft's crew.

The following day, the big truck limped into Alice Springs, stopping first at the telegraph station to 'give an account of ourselves to Mr Allchurch'. In reading the small mountain of mail waiting for him, Blakeley opened a letter from the Sydney office ordering him to hand over everything to Errol Coote, as he was now taking over to assist Lasseter in finding the reef.

The inevitable had finally happened – Blakeley had been sacked, the old-time bushman replaced by a journalist from Sydney who thought he was Charles Kingsford Smith.

Coote arrived by car to meet them and informed Blakeley that he was now in command, producing a telegram from Sydney certifying his authority. Blakeley showed the same respect to Coote's authority as the airman had shown to his. Blakeley let him have it. 'You seem to have been working fast; it is only about twelve days since I sent in word that we were returning,' he snarled. 'Anyway, that is the position you have been itching for all along, for not once have you complied with the company's agreement.'

'In what way?' Coote inquired.

'By not recognising me as leader.'

Blakeley was hugely resentful about the way everything had panned out. As far as he was concerned, he was simply another dupe in a long line of dupes taken in by the old tale of an El Dorado.

'When we got back to the Alice there were just as many people to welcome us as when we left,' he reflected bitterly. 'As we limped through the main street, three white men and two natives were all we saw; they did not even give us a second glance. The residents of Alice Springs are used to seeing expeditions come and go. They have had one or more a year for the past 35 years. Each crowd that comes has the same old maps, the same old dead man's tale, the same bags of gold specimens. Our turnout, the best equipped ever seen in this district, was a little different. We had the original man who had found his reef; but it cut no ice. It was the same old tale, only told in a different way.'

Blakeley had been instructed by the Sydney office to load the Thornycroft and the wreckage of the *Golden Quest* onto the train at Alice Springs and return to Sydney via Adelaide. The truck and the aircraft, the two technological marvels through which the expedition would return triumphant, were virtually scrap. 'We loaded the old Thornycroft with the remains of the *Golden Quest*, and what a miserable thing it looked, like a moth without wings.' The big truck that had been their home for two months was now nothing more than a rolling wreck, described as 'being in a pitiful condition: the big ends were loose and with the engine running they sounded as if they were likely to fly out of the crankcase at any moment, the cylinders were scored, and the crankshaft was only just hanging together'. Taylor drove the Thornycroft onto a railway flat car, where it was chained down; Blakeley recalled, 'One of the railway men, after tying it down and making it secure, stood back and said, "Well, there is the best load of experience I ever saw . . . Some people seem to get all the luck in the world. I envy you chaps."'

•

Coote was now officially in command of the expedition and busied himself taking notes and stocktaking supplies while Blakeley and Sutherland spent several days waiting in the Alice for the train. Blakeley would head back to Sydney and give the company directors what for. As far as the new expedition leader was concerned, he wasn't worth giving the time of day. Coote, now desperate for allies, approached young Taylor to ask if he would stay on as the aircraft's mechanic, to which he agreed. Coote's new-fashioned aerial expedition to locate Lasseter would not involve troublesome motor vehicles for ground support. Instead he decided on using camels – something Blakeley had been begging the Sydney office for weeks ago. Coote wired the company back in Sydney outlining his plan to fly the *Golden Quest II* from Alice Springs to Ayers Rock via the Hermannsburg Mission. The camel component would meanwhile march overland via Temple Bar, the Hermannsburg Mission and Running Waters, cutting through Bob Buck's property at Middleton Ponds and meeting up with Coote, who would fly to the rock. Coote found a suitable camel driver in Paddy Tucker, a part-Aboriginal guide who was able to put together a train of 15 camels at five shillings per head per week. Tucker had been recommended by a local missionary and would be paid two pounds ten shillings per week plus food and tobacco, the cameleer advising Coote it should take about 14 days to travel overland to the rock.

As incredible as it seems today, even in 1930 few white people had any idea what the great sandstone monolith Uluru – known by European Australians as Ayers Rock – looked like. Even though Coote knew of it, there were many giant, monolithic features spread throughout Centralia, and he wanted to know precisely what he would be aiming for on his flight out into the Never Never. Mr Kramer, the missionary who had recommended Tucker as the camel guide, sat down and drew several sketches of the rock for Coote, who commented later, 'According to him there was no mistaking the giant monolith, and he was right!'

Taylor asked Coote if he could travel with the camel train rather than fly as the passenger in the *Golden Quest II*. Coote said the young English mechanic was keen to 'get some experience of camel riding' and take photographs along the way. But perhaps Taylor, who had initially flown to Alice Springs with Coote from Parkes in New South Wales and witnessed the cataclysmic consequences of his flying abilities, opted for a less dangerous mode of travel.

Tucker's long camel caravan with Taylor on board slowly plodded its way out through Heavitree Gap for the journey to Ayers Rock, where, in two weeks' time, it would rendezvous with the aircraft. Meanwhile, Coote planned to fly to the Lutheran mission at Hermannsburg, primarily to engage the services of Paul Johns's native guide Rolfe, with whom he had been greatly impressed when he had met him at Ilpbilla. Taylor had mentioned that Rolfe and the other guide had left Johns at the old airstrip to walk back to Hermannsburg. Coote figured that as he knew both Johns and Lasseter, Rolfe would be of invaluable assistance in tracking them down. Coote took off from Alice Springs headed for the mission some 100 miles away, taking with him the Commonwealth medical officer Dr Kirkland, who had coincidentally been called out there to treat two injured men.

Apart from a typically Errol Coote close shave with the telegraph wires on takeoff, the 45-minute flight out to Hermannsburg went smoothly, the aircraft landing on the airstrip without a hitch. For the locals it was quite a spectacle. 'Hundreds of black children were lined along the aerodrome,' he recalled, 'and, as I gave the engine short bursts, taxiing to the mooring pegs they ran in all directions.'

They were welcomed by the mission's superintendent, Pastor Albrecht, and the local schoolmaster, Adolf Heinrich, both of whom oversaw this remote German-speaking religious outpost situated on a bend in the Finke River. These were catastrophic times. Not only were the missionaries enduring the drought, but the local Aboriginal community had been battling the worst incident of scurvy the mission had ever seen. Since the 1870s the Hermannsburg Mission had struggled to survive out in the dead heart despite drought, flood, disease and poor soil. Its sheer remoteness and the failure of Christianity to supplant tribal beliefs made its existence almost untenable. But the mission survived, finding its niche in the wilderness, especially during the violent war between Aborigines and pastoralists, when it became a kind of haven.

While Dr Kirkland attended his two patients, Heinrich and his wife entertained Coote, who eventually brought up the subject of hiring Rolfe. 'When I told him I wanted him to join our team and look for Lasseter he was quite willing to go,' remembered Coote. 'Heinrich also suggested another boy, a Pitjantjatjara native from the Petermanns, who was anxious to get back to his own country. I said that he could go along too.' The two Aboriginal boys were kitted up and sent on their way, headed for Bob Buck's property to join Taylor's camel party.

Dr Kirkland's work at the mission took longer than expected, and he and Coote returned to Alice Springs the next day. Fearing Coote had met with another disaster, search parties from Alice Springs were already out 'scouring the countryside looking for a wrecked machine', he recalled. The locals were not impressed when they finally turned up.

Keeping a low profile, all Coote could do was wait for the agreed two weeks to pass before taking off for Ayers Rock, but when the day arrived Alice Springs was being battered by a tremendous sand storm. When it was over, the dust simply 'hung suspended in the air for thousands of feet'. Takeoff would have to be the following day, and by midday the dust had finally dissipated enough for Coote to climb in behind the aircraft's controls. Ayers Rock would be a four-hour flight to the south-west. He had notified the Government Resident that he would fly via the Hermannsburg Mission on the way out and back, and if he was not back within a month they should send out a search party.

Soon after takeoff, the little biplane struck dangerous weather. 'A strange phenomenon was creeping over the desert brought closer by the headwind I was bucking,' wrote Coote. 'Away on the horizon the dust that had been over Alice Springs was coming back towards me. Here and there in front of it, I could see heavy thunderstorms. The lightning was flashing in a most startling fashion. Although the storms were in isolated patches, the lightning seemed to be all over the place.' Thankful at reaching Hermannsburg, Coote was advised by Pastor Albrecht to fly back to the mission after a week to report that they had successfully set up a base camp. Coote informed him he had told the Government Resident that he would report back in a month, to which Albrecht replied, 'A month is too long if anything happens to you, you would be dead in a month.'

'The compass course I set for the rock was south west,' wrote Coote. At this point, there were three separate parties heading in roughly the same direction: Coote piloting the *Golden Quest II*, Taylor with Tucker and his camels, and presumably Lasseter and Johns, who had been slogging along the track from Ilpbilla.

'I kept a sharp lookout in case I should see Taylor and the camel team,' recalled Coote, 'but not a moving object disturbed the peace of the sandy solitude . . . I was in a dead land.' Ahead of him was the vast salt-encrusted expanse of Lake Amadeus, where as he described, 'dotted here and there were small islands covered with dead-looking vegetation. The whole place seemed dead, excepting for occasional pools with water on the top. These were the danger spots. Beneath that crystalline surface of dried salt was a black, treacherous morass. Camels have rarely crossed this lake, but on every occasion when they have, they have become almost frenzied with fear. Snorting and screaming they miss their foothold and sink to their girths in the slimy, stinking filth that has been rotting for centuries under the white crust of salt.'

Eventually he found his objective. Coote's impression of seeing Ayers Rock and the Olgas for the first time, particularly from the air, clearly stuck with him: 'Soon I was flying over the narrow extremity of Lake Amadeus, a truly terrible place. The air was very bumpy here, and I had to hang onto the controls with my hands, feet and almost my teeth. The plane was being thrown everywhere. Then, like a red ghost, out of the dust haze came Ayers Rock. Square and stark it looked, and I knew it was the monolith I was searching for. The queer knobs of Mount Olga were different: they had definite shape and looked like a group of mosques set in the desert. Brilliant blue in colour, they were more picturesque.'

With an eye on the fuel gauge, Coote flew around and around the mighty monolith, scanning the ground for any sign of Taylor and the camels – nothing. He began to panic, wondering if Taylor had mistaken Mount Conner for Ayers Rock. Realising he only had about two hours of fuel left, he decided to make the 57-mile dash. But there was no sign of anyone at Mount Conner either, and he winged it back to Ayers Rock, fully aware that the fuel he had just used for his side-trip had ruined any chance of flying back to Hermannsburg. To the south-west of the rock he spotted a sandy patch on which to land, and came in with the engine at quarter-throttle. The wheels lightly touched on some porcupine grass, the aircraft slowing as soon as it was on the sand. Everything seemed to be going well and Coote closed the throttle.

Then the unexpected happened. Coote recalled, 'I had almost pulled up, with the engine just ticking over, when a sound like a pistol shot made my blood run cold.' The propeller had caught a four-inch stick, which smashed the edge of the blade. 'My position was now precarious in the extreme . . . I was alone in the desert hundreds of miles from civilization.' With a broken propeller and hardly any fuel, he was shaping up to become the first white man to die at Ayers Rock.

Coote had no idea what on earth had happened to the ground party. They'd had ample time to arrive at the rock – all sorts of scenarios ran through his head. Coote himself was in fact five days overdue. Maybe they had grown tired of waiting and gone back. Had they overshot their destination? Were they stuck somewhere? Were they prevented from getting there for some unforeseen reason? He knew he'd have to wait at least a week before a search party was sent out from Hermannsburg.

He took stock of the supplies he had, most of which had been given to him at the mission: a bottle of medicated wine, Horlicks malted milk tablets, Oxo beef cubes, tea, a flagon of water, 18 eggs, half a dozen carrots, two cabbages, two small fruit cakes and a bottle of Bovril. While circling the rock from above he had noticed some rock pools and now set off to find one, dragging a stick behind him so that he wouldn't lose his way back – a trick he had seen Lasseter employ when he went on his long walks. It was quite a hike to the rock from where the aircraft had landed, and he painfully discovered that his flying boots were not designed for slogging it out through the mulga. Soon his feet were chafed red raw. He found a rock hole, full from the recent rains, 'where the water was beautiful'. He now knew that when his flagon of water on board the aircraft ran out, he would still be able to find some. That night, Coote slept inside the aircraft's cramped cockpit.

He set out the next day determined to circle the rock, taking with him a tin of red aeroplane dope, used to repair torn fabric covering the fuselage, to paint signs on the rock. 'On reaching the rock I painted my first S.O.S. sign on a huge boulder... The legend read "S.O.S. Plane 5 miles S.W. Rock Coote 29/10/30".' In his wanderings he stumbled across the tyre tracks of the Michael Terry expedition, a discarded oil drum and a Lifebuoy soap wrapper – signs of civilisation compounding his fear of being alone.

He fully circumnavigated the rock on foot and eventually made his way back to the *Golden Quest II*. He later recalled, 'my task was done: I had painted the signs. There was nothing left to do now but wait at the plane for succour – or death.'

At night he would fire shots from his pistol, but no reply ever came. 'During the day I stayed in the shade, under the wings of the plane; the heat was terrific.' Coote decided that when the water in the flagon eventually ran out he would leave the plane and make for the waterhole he had discovered at the rock; and after eight days this time came. He bundled whatever he had into two loads, but was so weak that he decided to move it to the rock in relays. This meant tripling his journey in the ever-increasing heat. As he recalled, 'This ninth day was going to be a blast furnace occasion. And it was.'

Then, to his horror, Coote discovered that the waterhole he had found more than a week earlier had evaporated. 'I nearly collapsed as the thought now struck me that this had probably happened to all the other waterholes.' Frantic, he scooped up some of the earth and put it into his mouth, hoping to suck out some moisture, but there was none. His rapidly escalating fears were soon eased when he noticed dense undergrowth along the rock's base, perhaps signifying water. 'I tore through the bushes. A foot long black snake reared and hissed. Madly, I struck at it with a stick, and flung it several yards away. And there at my feet was clear blue water – thousands of gallons of it, deep and cool. I bent down, plunged my face in it, and drank and drank.' Coote had found a cave, a cool haven from the outside temperature that was now somewhere in the mid-forties. He was down to his last cigarette.

It suddenly occurred to him that he wasn't the only person suffering out here in the cruel dead heart. 'But how was Lasseter faring?' he wondered. 'Was he in a cave too? Or was he plodding along, monotonously swaying in elliptical circles to the noise of the camels' feet crunching through the sands of eternity?'

Coote's state of mind was faltering. 'Yes that was it: I was here waiting to die . . . Mentally I cursed Lasseter, cursed the gold, cursed Taylor – and most of all cursed the rock.' Under full moonlight he decided to press on to make for the north-eastern corner of the rock, as that was the point the camel party would presumably reach first.

He no longer fired the pistol at night. 'Ammunition was running low and the aboriginals might come in any day – and I would want what remained for emergencies . . . Ayers Rock, I had been warned, is sacred to the aboriginals, and the presence of white men there would be resented.'

Coote continued his journey under cover of darkness, each vast spur he passed followed by another – and then another. His thirst was starting to cloud his decisions. He knew there was a waterhole somewhere ahead, and peeping over the top of a ridge he spied 'a dozen fires, strung out in a line . . . Aboriginals. The thought struck me like a thunderbolt,' Coote recalled. Terrified he would be discovered and killed while he slept, he suffered a terrible night, and in the morning despaired at not having found water. 'By dawn I was desperate. It was impossible to last out any longer knowing that water was only a couple of hundred yards away . . . my thirst was now unbearable. So I made a break for it . . . keeping a watchful eye on where I had seen the fire being lit. Nothing stirred; no-one moved from that quarter. Perhaps they were already astir, hunting for me.' His thirst got the better of him. Coote was ready to sprint to the waterhole and suffer whatever consequences would result, when his world suddenly changed.

'At the end of the spur I nearly dropped with astonishment – there were camel pads, and there was a white man bending over a fire. It was Taylor!'

Taylor was equally shocked. 'Where the hell did you spring from?' he asked.

'Where the hell have you been?' Coote retorted. 'Give me a mug of tea and a hunk of damper. I haven't had a feed since the morning of the 29th – not a real feed.' Coote certainly had a story to tell, but what had happened with the ground party? 'You're late,' he remarked to Taylor, 'tell me *your* story while I attend to the great starve.'

In fact, Taylor's experience hadn't been much better than Coote's. He and his party had suffered long delays from wild thunderstorms on the way out to Bob Buck's station. The camels became bogged and threw their loads. Their A-frame packs had become waterlogged and there was nothing to do but wait for the straw-filled pads inside to fully dry out – a wait that took up to two weeks. They had been told by a fellow named Breaden at Buck's place how to find their way to Ayers Rock, but Taylor insisted on following the map, 'which like all other maps of the time was faulty', Coote observed. The party became seriously lost, winding up at Mount Conner instead, 60 miles to the east. With very little water left, they finally made their way to the rock, found a waterhole and set up camp. A two-week journey had taken a month.

That afternoon, Taylor and Coote rode out to the *Golden Quest II*. The young mechanic removed the propeller and repaired it by riveting a section cut from a petrol tin over the broken part of the blade. It was a remarkably simple repair, and an effective one, Coote successfully making a brief test flight. But on his return, he discovered Taylor in considerable pain, having been kicked in the chest by a bolting camel. For an exhausted Taylor, this was only the beginning of a series of painful complaints.

Coote prepared to take off and return to Alice Springs. His intention was to fly the aircraft to Adelaide for further repairs and then fly back to Ayers Rock to re-join Taylor. He advised the young mechanic to wait 12 days. If he hadn't returned by then he should make for the Hermannsburg Mission via Buck's station.

Coote fired up the aircraft's engine and lowered his flying goggles onto his face. 'Well, so long,' he said to Taylor. 'I'll be seeing you later; then we'll get on with the job. Everything should be plain sailing now.'

But as with everything else on the CAGE expedition, nothing would be plain sailing. On landing at the aerodrome at Alice Springs, Coote found an agitated Government Resident waiting for him. All hell had broken loose when he had failed to return as scheduled. The Hermannsburg Mission had begun sending out SOS messages, search parties riding teams of camels had set off from Alice Springs, and the Royal Australian Air Force had dispatched three aircraft from Melbourne: at that very moment they were in the air on their way to Alice Springs. Rarely one for taking responsibility, Coote – the expedition leader – apportioned the blame: 'The Mission altered the arrangements' and 'the ground party took a month to do the trip. They were a fortnight late reaching the rock.' The Government Resident coldly informed him, 'I'll have to wire the Air Board to stop those planes.'

The directors in Sydney were apoplectic. Coote's expedition had been bungled from the start, worrying shareholders and throwing unwanted light on the operation, arousing the press.

Coote received blunt instructions to fly back out to Ayers Rock, pick up Taylor and return immediately to Sydney. His time in the Red Centre had been terminated. Like Blakeley before him, he had been relieved of command, and according to Coote, as far as the Sydney office was concerned 'Lasseter was to be left to his own devices'.

The Gipsy Moth with its bodged-up propeller repair was hardly airworthy enough to make the risky flight out to Ayers Rock, let alone the long haul to Sydney. Instead, Coote decided to try to fly the 900-mile distance to Adelaide to have a new propeller fitted, and then fly back to pick up Taylor. But 30 miles out from the Alice, he noticed there was 'something seriously wrong with the plane' and then struck strong headwinds. With the aircraft's engine running erratically, Coote decided to head back. He would send the propeller to Adelaide by train. Nobody was bringing him back to Sydney yet. Not while Lasseter was still out there looking for his gold.

29

THE SYDNEY CROWD

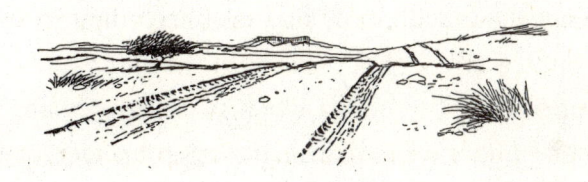

In the bleak financial depression that gripped Australia in 1930, an irascible Sydney journalist named Lennie Lower wrote a humorous novel titled *Here's Luck*, which took Australia by storm. It regaled readers with the hilarious everyday ups and downs of the hapless Gudgeon family as they battled through life during the Great Depression. 'Here's luck' was the familiar toast drinkers spluttered before throwing down the last mouthful of beer to stagger home from a dim, tiled-wall pub closing its doors at six o'clock. In the grim depression days of 1930, luck was the only thing people could hope for. But it was the one thing the CAGE expedition never had.

Fred Blakeley arrived back in Sydney, as he put it, 'unannounced and very much unwanted'. He fronted the company's board of directors and 'gave them a verbal report of the whole trip, sparing nothing and nobody. I just told them exactly how things stood and of Lasseter's total failure. There was not one redeeming feature in the whole thing.' Further, he warned them they would not see Lasseter again, advising the meeting, 'It was no use looking for Lasseter as Lasseter would take good care not to be found.'

When Blakeley had finished, the chairman John Bailey simply said, 'Well, we have heard Blakeley's report, that's what we came here for.' The meeting broke up. 'I was amazed at the silence of everyone,' Blakeley recalled. It was as though the Lasseter phenomenon had grown to become almost a religious cult of blind acolytes – and Blakeley had taken the dangerous decision to question it.

He was waiting by the lift when two of the board members approached him, one putting his fist up to his face, saying, 'You and Colson won't ever get that reef. We all know about you two. You think now that you have got everybody out of the road and Lasseter discredited, you two will go out and find the reef, but we are sticking with Lasseter and we hope when you try and double-cross us that you will perish of thirst.'

'I left in disgust,' wrote Blakeley.

Blakeley had furnished Ern Bailey, the company secretary, with his written report to be read out at the next general meeting, but still had another task to complete while in Sydney: to return the almanac loaned to Captain Blakiston-Houston by Rear Admiral Evans. He telephoned the RAN flagship HMAS *Australia* and spoke to the admiral directly, who insisted Blakeley should 'not let the book out of my hands until he took it from me'. He was invited on board the following day, where the grateful admiral received his prized book and feted the 'Leader of the Lasseter Expedition' among the ship's officers. Blakeley was invited to attend a luncheon on board the following week – an event that greatly impressed him – and upon leaving apologised to the admiral. 'I was sorry to say that there would be no big reef, as Lasseter was only chasing a dream.'

Blakeley attended the next CAGE company general meeting and sat waiting for his report to be read out. The company chairman read aloud 'something he said was "extracted" from my report. When he finished there was silence such as might be found in a Sunday school class. Without stopping for questions he went on to other matters . . . The meeting came to an end and I was clean-bowled . . .'

Whatever Blakeley said or wrote about the collapse of the expedition didn't make one iota of difference to Jack Bailey. As far as he was concerned, the old overlanding cyclist was just a waste of time and space.

Blakeley wasn't to know that Bailey had an ace up his sleeve: the letter Lasseter wrote in invisible ink revealing the coordinates of his reef, still locked in the vault of the Bank of Australasia in Sydney. The longer Lasseter was alone out in the wilderness, the more chance he'd die – the legal stipulation whereby Bailey could get his hands on the letter.

30

THE STRANGER IN TOWN

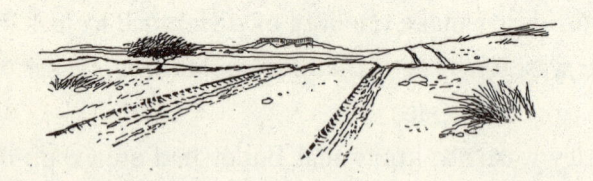

There was some sort of commotion brewing in Alice Springs. Word quickly flashed around town that a lone, dishevelled stranger had wandered in with five exhausted camels. He had said he had something to do with the search for the lost gold reef with that fellow Lasseter.

It was Paul Johns – as everyone described, very much alone – and looking the worse for wear. Filthy, drained and penniless, an exhausted Johns had finally made it back to civilisation with his worn-out camels. It had been two months since he and Lasseter parted from Blakeley and the expedition to find the gold reef, but only Johns had returned to Alice Springs.

The now cashiered Errol Coote, who had been hanging around the Alice while the aircraft's propeller was being repaired in Adelaide, was shocked to learn of Johns's sudden emergence from the desert and was desperate to know if he had news of the reef, and of Lasseter. According to Coote, the young German turned up at his tent, and 'obviously under strain of suppressed excitement, told me a fantastic story'. After parting from the expedition, he and Lasseter had backtracked to Mount Udor and turned south. 'They crossed Lake Amadeus, called in at Ayers Rock and then went via the Petermanns into wild broken mountainous country.' Johns said that Lasseter had left him to stay with the camels for a few days and went on alone to find his reef, returning later with bags of gold specimens. Lasseter wouldn't show them to Johns . . . there was an argument . . . Johns drew his revolver . . . there was a fight . . . 'At this point Johns became very excited,' recalled Coote, who told him to 'cool down'. In the end, said Johns, he and Lasseter agreed 'to let bygones be bygones' and go their separate ways, the prospector giving Johns a letter to hand to the Government Resident in Alice Springs. Johns produced the sealed envelope, curiously mentioning that the fight was discussed in the letter: 'He has reported me in it yet he wants me to go out there again.'

Coote had only encountered Johns once before, when he landed the *Golden Quest II* at Ilpbilla with Pat Hall, and from the very outset was suspicious of the strange young dogger. Coote recalled the conversation.

'Anyhow, how do you know what is in the letter? Did you open it?' asked Coote.

'Yes of course I did. You don't think I'd let him get away with anything that might not be correct do you?'

'Well, you had better trot it over to the Government Resident as quickly as you can,' Coote advised him. Something didn't seem right.

Coote followed Johns to Vic Carrington's house, where the young German presented the Government Resident with the envelope containing two handwritten letters from Lasseter. Johns informed him that the prospector was still out in the desert in search of his reef. According to Coote, the contents of the letters were startling, Lasseter claiming he had rediscovered his reef, but it was 'not quite as rich as I thought it was'. The message said he had pegged out six gold leases and wanted Johns to return with fresh camels and supplies. He would wait a few weeks at Ilpbilla for Johns to return, and after that he would travel 60 miles south-west from the old airstrip and wait at a waterhole. If Johns failed to turn up, he would move across the Western Australian border to Lake Christopher to meet a man named Johansen, from Boulder City, near Kalgoorlie. Who was Johansen? No-one had the faintest idea.

The other letter, addressed to the 'Head of Police', took an even stranger turn. It detailed the incident when Johns and Lasseter had been involved in some sort of brawl, the prospector demanding the young German be placed 'under arrest and jailed' for threatening him with a firearm. Carrington put the dingo scalper on the spot. 'What the hell did you do to Lasseter? He said you had a fight and nearly murdered him. Are you sure you didn't do him in?' Johns said later he assumed Carrington wasn't serious about him killing Lasseter. Nevertheless, word soon spread around Alice Springs that the lone dogger had indeed murdered him.

There were reasonable grounds for suspecting Lasseter had been 'done in'. At that precise moment the nation was captivated by one of the most sensational crimes to emerge from the Never Never. A year earlier, Snowy Rowles, a stockman working along the rabbit-proof fence in Western Australia, had murdered three men, and disposed of their bodies in a manner described in a detective novel written by a co-worker, Arthur Upfield. An emerging published author struggling with his latest plot, Upfield had asked his workmates on the fence to help him devise a method of eradicating any trace of a dead body so that there was no corpse for his detective hero to find. It had been a problem for Upfield's story, discussed regularly around the campfire with a young Snowy Rowles quietly listening in. A colleague suggested burning the body with that of a large animal, sifting any metal fragments such as tooth fillings from the ashes, dissolving them in acid and grinding any bones left to dust. Snowy Rowles took all of this on board. Over a period of months, three men – James Ryan, George Lloyd and Louis Carron – were poisoned with strychnine and disposed of in precisely this way. Rowles emerged from the desert alone, driving a truck belonging to one of his victims and attempting to cash cheques from another. Arthur Upfield himself was suspected of the murders until police recognised Snowy Rowles as the escaped prisoner John Thomas Smith. The trial was described as 'the most comprehensive in Western Australia's history'. Rowles – or Smith – was hanged in 1932.

Whatever anyone suspected Johns had done, he nonetheless convinced Carrington and police sergeant Littlejohn that the idea he had 'done him in' was out of the question. From the letters he had delivered, it was fair enough to assume that Lasseter must still be alive out there somewhere. Yet the dogger's return to Alice Springs to turn over Lasseter's damning letter about him to the Government Resident had been, in a word, leisurely. He had clearly been in no rush to go back out to assist the idiosyncratic and impetuous prospector, stumbling alone in the desert.

So what exactly had gone on when Lasseter and Johns had struck out on their own two months earlier? Precisely what transpired between the pair in the wilderness late in 1930 will never be fully known, as Johns was the only witness. In the years that followed, Johns agreed to a few rare interviews with newspapers about his and Lasseter's unfortunate journey into the Petermanns, and in 1932 he gave a verbal statement to author and journalist Ernestine Hill.

Today, some of these newspaper accounts seem sensationally overwritten, no doubt embroidered by enthusiastic journalists ghost-writing for the young German who had arrived in Australia in 1926 with virtually no understanding of English. Other accounts, such as his statement to Ernestine Hill, are vague and somewhat melancholy, conveniently avoiding his questionable role in the Lasseter mystery.

•

Lasseter and Johns's journey to find the reef in September of 1930 was an unquestionably harsh one. Johns's camels were already in a fairly parlous state when they set off from Ilpbilla. The newly arrived young European and the 50-year-old prospector whose claimed experience in the region was at best doubtful were heading off into a Centralian inferno with no native guides. The

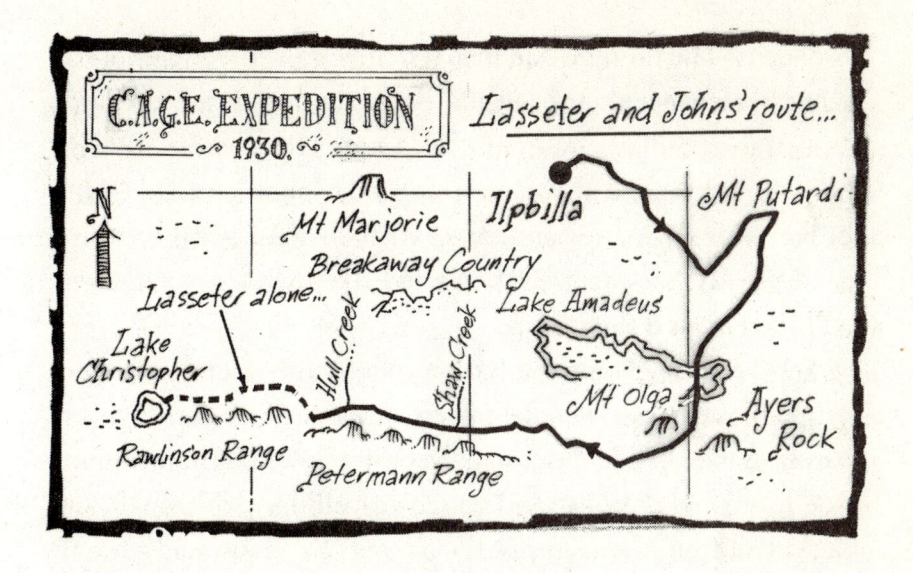

> **C.A.G.E. EXPEDITION 1930.** — *Lasseter and Johns' route...*
>
> Mt Marjorie · Ilpbilla · Mt Putardi
> Breakaway Country
> Lasseter alone... · Hull Creek · Shaw Creek · Lake Amadeus
> Lake Christopher · Mt Olga · Ayers Rock
> Rawlinson Range · Petermann Range

plan was to approach the Petermanns in a giant arc sweeping back east, turning south and then west. The route they would follow was slightly convoluted in order to link up with waterholes Johns knew of. But the savage drought would seriously upset their plans.

Setting off, they ambled back along the now well-worn track from Ilpbilla before heading south-east towards a known waterhole where, to their shock, the spring they were counting on was bone dry. Taking a gamble, they headed north-east towards Mount Peculiar, where Johns was hopeful they would find water, and this time their luck came good. Here they began to divest the camels of some of their load, burying 150 pounds of flour as a food dump. If the return journey was problematic, they could join the dots between depots. On a map, this detour certainly takes them a long way from their proposed destination, but as they were learning, water was becoming increasingly scarce. It would be the last time they would strike water for a long time, as they were venturing into serious desert country.

The camel train wandered roughly 80 miles south towards a narrow 10-kilometre crossing of the blinding white salt-pan Lake

Amadeus. The drought had inadvertently proven a godsend: had there been water there it would have stopped their expedition in its tracks. Still, the salt crust surface was treacherous. 'We got halfway across and the camels bogged,' Johns recalled. 'They broke their nose-lines and chucked their loads, and were too tired to extricate themselves. We fed them while freeing them, and then chased them back.'

In 12 days of crossing the desert country they found no trace of drinking water, not even for camels, and their own supplies were barely usable. 'We had two gallons of not water but hot steaming smelly stuff – the heat and rough travelling had brought out the petrol from the seams of the drums.' Food too was not only uninviting but also strictly rationed. 'We were living on tinned salt beef and rice, but were mostly too weary and disheartened to eat – especially Lasseter,' said Johns of their miserable existence. 'Millions of black ants crawled over us at night.' Before them lay countless sandhills, which took days to traverse – the camels noticeably deteriorating. If they perished – to be stripped to the bone by the desert ants – then so too would Lasseter and Johns. Everything hinged on the survival of the camels.

They would soon encounter two extraordinary natural wonders: the giant red-domed rock formations known as the Olgas and the breathtaking sandstone monolith Ayers Rock. At Mount Olga they were ecstatic on finding a freshwater spring; Johns recalled, 'The camels slunk down silently, and put their heads in the hot sand. When we took them to the spring they wouldn't drink. I dived my head in the water and nearly cried. I then drew bucketfuls of the water and gave each camel a few sips. At sundown they filled themselves.'

They pressed on, heading directly west. At Mount Stephenson they stumbled onto 'a big reef of crystallised quartz', but no gold. Johns had his doubts about what they would find.

'Lasseter knew all this, and yet he did not impress me as a man who knew the country, but rather as one who had read about it,' Johns commented in his statement to Ernestine Hill. The further they travelled, the less cooperative Lasseter became; Johns described him as 'very reticent, a silent man, and would often ride without speaking'. The propensity for paranoia he had displayed with the CAGE expedition members had now shifted squarely onto Johns.

As the journey continued, Lasseter's suspicion of Johns grew even more intense. After all, he was wandering out here alone with some gun-for-hire Fred Blakeley had teed up for him. What was to say that when they found the reef, this strange young dingo scalper – with his arsenal of guns and strychnine and boning knives – ensured Lasseter somehow disappeared? If Johns – whose daily work was killing and dismembering animals – did his work properly, then no-one would ever find Lasseter's body. The camel train plodded silently on.

In his statement, Johns said they passed through the Petermanns and crossed the border from the Northern Territory into Western Australia to reach the Rawlinson Range. 'I think Lasseter thought we were nearing the end of our journey,' said Johns. 'At any rate we were nearing the end of our rations.'

On reaching a waterhole, Lasseter announced to Johns that he would travel alone from here on, saying he was convinced the reef was fairly close by. He would be gone only two or three days on a brief reconnaissance mission, and instructed Johns to stay put. Frustrated, Johns had to resign himself to waiting idly with the camels while his erratically motivated boss wandered off by himself. This was risky stuff – two men together had a better chance of survival than being apart. If Lasseter became lost or either of them came under attack . . . it didn't bear thinking about.

Exactly how long Lasseter was gone isn't precisely known – somewhere up to five days perhaps. But return to the camp he did – triumphantly.

Lasseter announced to Johns that he had relocated his reef, and to prove it he brought with him a bag of gold specimens. For Johns it was nothing short of miraculous. Here he was – he had gambled on following the wily old prospector out into the Never Never and it had paid off. Blakeley and Coote and all the others had given up on Lasseter, but not Johns. He'd been canny enough to stick with him, and now he would be present at the unveiling to the world of the greatest gold find in history. He knew Lasseter's fabulous story only too well – how Harding the surveyor all those years ago had said Lasseter's gold specimens were the best he had ever seen – and now it was Johns's turn to see the treasure this El Dorado held. He waited for Lasseter to show him his find – to produce the fabulous nuggets of gold brought back from his golden highway. But there was no hint of any dazzling revelation.

Johns asked Lasseter to see the specimens inside the bag, but the prospector refused, telling him it was 'none of his business'. Lasseter's response floored him. The young German had spent weeks guiding this difficult and eccentric fellow through the worst country imaginable, and now not only had he denied him the privilege of accompanying him to see the fabled reef, he wouldn't even show him the spoils. Johns's demeanour suddenly changed – he smelled a fraud.

The realisation that Lasseter was some sort of impostor would have brought his blood instantly to the boil. Only a few months earlier while working at the Hermannsburg Mission he had been the victim of a con man – an itinerant German 'nobleman' who titled himself Hauptmann von Berlichingen. This 'Hauptmann' – or captain – somehow arrived unannounced at the mission, where he was feted and treated with much fanfare as a visiting celebrity. He was even given Johns's bed. Yet, for all the to-do about Hauptmann von Berlichingen, something didn't seem right. In time Johns grew suspicious of the 'captain', and while returning from Alice Springs with the mission's mailbag, he opened all of von Berlichingen's letters, read them, resealed the envelopes and placed them back in the Alice Springs post box. A German friend at the time described Johns's impropriety in opening the letters as 'a scandalous breach of trust'. Johns continued rifling through the visitor's mail many times, and discovered that von Berlichingen was indeed a fraud: he had made a bet back in Germany that he could travel through Australia without paying one penny. Furious at being taken advantage of, Johns tipped all the opened mail out in front of von Berlichingen, shouting, 'Here you are old man! Here's your mail. You have been an impostor and a liar and a despicable sort of person because you can't be truthful. Is there anything you have to say about it?' Von Berlichingen hastily left the mission, and Australia, sailing back to Germany.

Now running low on provisions, it was decided they should turn back for the supplies stacked at Ilpbilla. By now relations between the pair had degenerated into an impasse. Johns had made up his mind Lasseter had taken advantage of him and he wasn't letting him get away with it. Interestingly, the German con man Johns had encountered at the mission had described himself as 'a wanderer, a dreamer', and it looked for all the world to Johns that he was saddled with another.

Having crossed Lake Amadeus, they reached a food dump Lasseter had left on the way out. The pair sat solemnly, eating fruit from one of the recovered tins. The angst brewing between the two since Lasseter had refused to reveal his gold specimens was compounding by the minute. Johns's impetuous nature would ensure that sooner or later he would say something that would push whatever little patience Lasseter possessed to the limit, or beyond it.

Johns figured he'd had enough. He was ready to inflame the situation to bring on a confrontation, just as he had with von Berlichingen. 'Why did you dump all this stuff here at this impossible place? You had no reason at all for it,' Johns baited Lasseter. And he knew just how to get a bite from the prospector. 'But I think I can tell you why. You wanted to mark time; while you are here, your lawyer in Sydney collects 10 weekly from the company, therefore, the longer this trip lasts, the better for you.' Lasseter sat silently, glowering while he mopped up the tinned fruit from his enamel plate. Johns then called Lasseter 'a liar'.

It was the flashpoint that within a second could have easily resulted in a murder, then and there. Outraged, Lasseter stood up and flung his steel dinner plate at Johns, striking him in the face. The prospector sized up to Johns, shouting at him to put up his fists, whereupon Johns drew his revolver, cocking the hammer and aiming it at Lasseter, and shouted back, 'The only thing I'll fight with is this.'

Lasseter launched himself at the young dogger and the pair rolled around on the stony earth, both grappling for the gun. Johns was taken aback by just how powerful Lasseter was, and was soon losing the upper hand in the melee. The fight for control of the weapon turned particularly vicious, and the trigger was pulled, Lasseter's thumb caught between the hammer and the revolver's cylinder. The pain was tremendous, the firing pin on the hammer piercing Lasseter's skin and tearing a sinew. Lasseter wrenched the pistol from Johns and trained it on him, neither knowing what would happen next. Johns froze, glaring at the revolver's muzzle now aimed squarely at him, Lasseter's bleeding hand trembling as it held the weight of the gun.

That moment, when the two men faced off at gunpoint, seemed to last for hours, and when the reality – and the gravity – of what had just happened sank in, Lasseter hurled the pistol into the scrub.

That was the end of the expedition. Johns said they agreed to 'let bygones be bygones', but their fractious partnership was only to last until they reached Ilpbilla.

'We travelled back in ten days,' said Johns. 'At Ilpbilla the camels were skin and bone. I told him his best plan was to wait at Ilpbilla while I went in to the Alice for fresh camels, to return in about three weeks. Lasseter replied that he was going on his own with the remaining camels and that I could go.'

Lasseter gave Johns the letter to give to the Government Resident. 'He stood and watched me go,' recalled Albert Paul Johns – the last white man to see Harold Bell Lasseter alive.

31

THE SEARCH FOR THE SEARCHERS

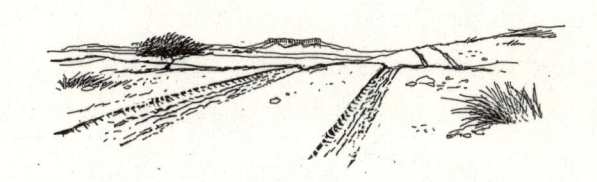

Word got back to the directors in Sydney that some German fellow whom Fred Blakeley had signed on had turned up in Alice Springs saying he'd left Lasseter alone in the desert somewhere. The Central Australian Gold Exploration Company was now in an invidious situation. Not only were they looking for a lost reef, but they had now lost the only person who knew where it was. So who was running the show out there? They had sacked the original expedition leader, Fred Blakeley; they had then sacked his replacement, Errol Coote. They were quickly running out of prospective leaders on hand in Alice Springs. There was really only one man still standing: the 22-year-old English aircraft mechanic Phil Taylor.

Taylor had returned from Ayers Rock after Coote's aerial expedition had been scrapped, arriving in the Alice in a poor state. After Coote had left the rock, a willy-willy had dispersed Taylor's campfire and destroyed the party's supplies. He and his Aboriginal guides filled an empty fuel drum with water, lashed it to a camel and set off for Bob Buck's station. Taylor arrived badly dehydrated, after unwisely throwing away the petrol-tasting water.

He was wired by the Sydney office that the *Golden Quest II*, now back in action after a brief overhaul, was heading out to Alice Springs. The aircraft was flown by a new pilot, Captain W. L. Pittendrigh, accompanied by mining engineer S. J. Hamre, the company wanting them to fly to Ilpbilla to make an aerial search for Lasseter around the last place he had been seen. Taylor was instructed to take a load of fuel and re-establish a base there, rendezvous with the aircraft and assist as the ground party. The company also re-engaged Paul Johns. After all, he was the one with whom Lasseter had made an arrangement to return. But perhaps there was another reason for including Johns. Blakeley believed the company re-employed him 'for fear he knew too much'. Taylor and Johns with four Aborigines arrived at Ilpbilla with a team of camels a few days before the aircraft was due to take off from Alice Springs.

Pittendrigh and Hamre left Alice Springs aboard the *Golden Quest II* on 20 December, heading on a route that would take them north of the McDonnell Ranges. They had with them 30 gallons of fuel, six gallons of water, enough rations for seven days, maps and matches. The flight to Ilpbilla would be short.

At the re-established airstrip Taylor and Johns waited anxiously for the sound of an approaching aircraft's engine, but they heard nothing. As time passed, the pair grew more nervous. Eventually Taylor sent Johns on a desperate ride back to Hermannsburg, where the alarm was raised, with word coming back that the *Golden Quest II* had in fact gone missing. Messages were sent out by the mission's pedal-powered wireless to Alice Springs and on to Sydney, where the company directors made an appeal to the Minister for Defence.

On New Year's Day 1931, two Royal Australian Air Force Moth biplanes took off from Point Cook in Victoria, piloted by Flight Lieutenant Charles Eaton – regarded as the hero of the *Kookaburra* search – and 'Jerry' Gerrand, headed for the Territory. They took off from Hermannsburg Mission on 3 January, having interviewed Johns and Rolfe, and landed at Ilpbilla, where they met Taylor. In Eaton's report he states, 'Natives had told him that a plane was seen about 8 miles north of Ilpbilla on 20/12/30.'

Eaton came to the conclusion that the aircraft had turned around and was most probably located closer to Alice Springs. Observation flights were made one after another until, on the seventh mission, Gerrand's aircraft struck a concealed stump while taxiing, breaking a rear spar, rendering his aircraft unserviceable. The extreme weather had taken its toll on the RAAF Moths, Gerrand 'finding many parts affected by shrinkage and heat'. Two more RAAF Moths then arrived, piloted by Flying Officers Dalton and Evans, and the scale of the search enlarged dramatically.

On the ninth mission, the newly arrived Dalton thought he saw what looked like a large 'T' shape on the ground; even though he suspected it was a signal made by some of the mission boys, further discussion at Ilpbilla that evening saw the RAAF pilots concluding it could well be from the airmen. Eaton's report describes their discovery: 'Tenth search flight took off from Ilpbilla at 3 pm, flying direct to Dashwood Creek where the missing airmen were found, and food dropped from a parachute, together with message instructing them to remain there until picked up by motor car.'

It had been another CAGE debacle. Pittendrigh and Hamre had completely missed Ilpbilla and had swung the *Golden Quest II* in a wide circle around the Ehrenberg Range. On reaching the area around the Dashwood, they were forced to land. They had run out of fuel and it looked as though they would soon run out of water. But the gods smiled briefly on the airmen, and only hours later they were hit by a thunderstorm, quick-thinking Hamre ripping off his singlet, spreading it on the ground and soaking up the rainwater. He wrung it out in a billy.

They eventually found a waterhole along a creek, where they decided to set up camp. As the days passed and their supplies dwindled, they resorted to cooking lizards they had caught, but drew the line at some of the inedible native vegetation. Occasionally they heard the sound of aircraft but could never seem to get a fire going in time, and feared that eventually their rescuers would give up. They decided to scrape a large letter 'T' in the red sand with their hands, but Pittendrigh was not convinced it stood out enough. He filled it with white ash from their campfire. It was this white 'T' on the deep red sand that Dalton spotted from the air.

The story threw another unwanted spotlight onto the CAGE venture. Some thought it might have been a Coffee Royal-style publicity stunt. Others were wising up to just how much all of this was costing the taxpayer. Fred Blakeley was particularly critical: 'Big headlines appeared in the newspapers every day. In the anxiety about the lost airmen poor old Harry Lasseter had to stay very quietly "lost" until the limelight was turned in his direction again.'

Indeed, the search for the missing airmen caused four RAAF aircraft to be flown some 1300 miles from Point Cook; and Errol Coote's brief disappearance at Ayers Rock also saw air force planes dispatched, only to be turned around when he reappeared at Alice Springs. But no RAAF Moths for Lasseter, who was still wandering out in the desert alone.

Pittendrigh and Hamre had been extraordinarily lucky. Eaton wrote in his report: 'if it had not been for the phenomenal rain which occurred within a few hours after they drank the last water which they carried, undoubtedly both men would have perished.'

But it was not only the airmen who were close to perishing. Phil Taylor was suddenly in a seriously bad way. All those months of existing in the drought-stricken mulga, the pressures put on him in keeping the Thornycroft going, the dehydration, the nutritionless bushman's diet and the interminable rocking of an endless camel ride had wreaked havoc on his body, and in particular his kidneys. The excruciating pain shooting across his back grew worse and Taylor plunged into delirium. 'I never thought I'd get him back alive,' recalled Paul Johns, who tied Taylor onto his camel. 'I headed straight east for Hermannsburg.'

Pastor Albrecht, the mission's superintendent, had taken charge of Taylor's deteriorating health when he received a desperate telegraph message from the CAGE company in Sydney pleading for assistance to find Lasseter. After a brief discussion, Albrecht and the schoolmaster Heinrich decided the experienced Central Australian cattleman Bob Buck would be the only man for the search. They sent the young Aboriginal stationhand Rolfe to Middleton Ponds, Bob Buck's property, and fortunately he found the bushman at home.

The pressure was on to locate the old prospector. But as each day passed, hope of finding Lasseter alive was diminishing. The gold exploration expedition had become a rescue mission. Now the perennial question as to whether Lasseter's gold reef really did exist was transposed onto the prospector himself – did Harold Bell Lasseter himself still exist?

32

BOB BUCK

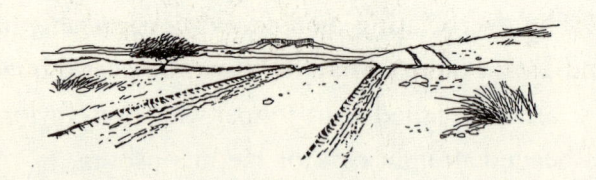

The fourth anointed expedition leader was the first who actually had some first-hand experience of the area where Lasseter and his gold reef were supposed to be. The wizened and weather-beaten Central Australian cattleman Robert Henry Buck looked and sounded and smelled as if the dead heart itself had erupted and given birth to him. It was obvious he was no old-time over-landing cyclist, or a reporter with an entry-level pilot's licence, or a 22-year-old mechanic from the Old Dart. Bob Buck, with his Henry Lawson moustache, his briar pipe and rabbit-fur felt hat, was outback Australia down to his elastic-sided boots. Even his name suggested he was some sort of antipodean Davy Crockett or Daniel Boone. Buck, as everyone knew, had been hired to clear and build the successful makeshift aerodrome at Ilpbilla and it's a wonder the CAGE company hadn't engaged his services earlier.

Buck knew how to find Lasseter all right – but his expertise came at a price. The drought had cost him thousands of pounds worth of cattle, and this sudden and urgent engagement to find

Lasseter and his reef would no doubt help in fending off the debt collectors. On receiving the request for help, Buck travelled through the night to reach Hermannsburg Mission, where the residents were amazed to see him so soon.

Hilda Heinrich, the wife of the local schoolteacher, recalled, 'We told him what had gone wrong and where he was likely to find Lasseter. But he was not inclined to go on a dangerous errand without his special stimulant, a case of brandy which took four days to turn up from Alice Springs.' Another frustrating delay, but once properly fortified, Buck was on his way.

Mrs Heinrich's recollections of what transpired after Buck left the mission were clear: 'When many weeks had gone by – and he didn't turn up . . . and he didn't turn up . . . our expectations eventually turned to doubt. We felt concerned. I don't know how long he was away. But lo and behold, a shout went up from the native camp one day: "Bob Buck is coming!" [It was 24 April.] And the first thing he did when he opened the door to my husband's study was to display a set of dentures. "I got him," he said laughing. I'll never forget those words.'

Buck told them he had found Lasseter's decomposed body in a shallow grave at Irving Creek, between the Petermann Ranges and Mount Olga.

Aside from being the local schoolmaster, Hilda Heinrich's husband was acting coroner at the mission. He immediately sent a report to the police in Alice Springs requesting advice as to whether he was required to go out and inspect the body. But as Mrs Heinrich pointed out in a 1969 interview with Lutheran pastor Philipp Scherer, her husband 'wouldn't have been able to recognise whether the body was that of Lasseter or anyone else. He had never seen or met the man.'

Buck, who had been in the wilds for 11 weeks, had also brought back other items presumably belonging to Lasseter: some photographs and letters found on the body. This was enough to satisfy the acting coroner that these items did indeed belong to the missing prospector. As for heading back out to sight the corpse, Mrs Heinrich recalled, 'The answer from the police was: No, it would not be necessary to view the body. If my husband was convinced beyond doubt that it was Lasseter, he could issue a burial certificate. So that is what happened. And on the burial certificate he stated he was convinced Lasseter had died of starvation.'

The questionable decision to not go out to formally identify or retrieve the body Buck had located has been debated ever since. And Buck himself didn't help the situation, as Mrs Heinrich recalled: 'And now, something I shall never forget. My husband had to take a sworn statement from Buck, and this had to be done on the Bible ... but I think Bob was really frightened when he held that Bible in his hand and had to swear on oath that he was about to say the truth, the whole truth, and nothing but the truth.'

Mrs Heinrich also alluded to subsequent rumours that perhaps the body for which her husband signed the burial certificate might not have been that of Harold Lasseter. 'I personally have no difficulty, on the grounds of circumstantial evidence, to accept that the body found by Buck was that of Lasseter – not withstanding the fact that the body was never properly identified. But, then, whose else could it have been?'

•

'LASSETER IS DEAD', ran the headline on the front page of the *Sydney Mirror* dated 29 April 1931. 'Starvation Ends his Search for Fabled Gold Reef – Body in Shallow Grave.'

The nation was in shock. The man whose gold find was to break Australia from the grip of a worldwide depression – the nation's saviour – had been found dead in a shallow grave in the middle of absolutely nowhere. Readers began to take stock of what had happened: the grand expedition that simply imploded, the monstrous expense in searching for inept airmen, and now this terrible, terrible news.

Bob Buck's discovery of Harold Lasseter's body turned the Centralian cattleman into an overnight celebrity; for everyday Australians trying to make sense of the Lasseter story, Buck's arrival seemed to reassure everyone that it was salt-of-the-earth bushmen who'd get the job done. He was the grounding force who brought sense to an otherwise illogical and inexplicable turn of events.

Buck's journey to find Lasseter after leaving Hermannsburg Mission was gruelling. In early February 1931 he set off with his train of 18 camels and Aboriginal trackers into the thick of a Central Australian summer. Under instructions from Taylor, his first objective was to reach Ilpbilla, the airstrip he had cleared for the Mackay Aerial Survey Expedition, in case Lasseter had returned there for supplies. But even though some of the food stores had been tampered with, there was no indication Lasseter had been there. From Ilpbilla they travelled directly for the Petermann Ranges. The camels became bogged while crossing Lake Amadeus, forcing everyone to manually carry their cargo to hard ground, where they were forced to camp. The fierce heat was particularly hard on the camels, who struggled to continue under their heavy loads. Some of the animals ate poison bush, and one nearly died as a result.

Buck's group then reached a place known as Puta Puta, near Shaw Creek on the northern edge of the Petermanns. Buck had brought with him some of his best workers, including a boy named Lion, and Billy Buttons, whom Fred Blakeley had once met by chance when returning with Lasseter from Ilpbilla. Buttons later recalled events that unfolded at Puta Puta.

'There we saw one old blackfellow, an old woman and a boy – a couple of kids. We asked the old man: "Did you see this fellow, one white man who has been out here?"'

'"No," he said, "I haven't seen him."'

'He knew about him all right, but he was afraid to tell, you see. Anyhow, we gave him and the others a feed. This boy Lion, who came from that country, could speak his language. Talking to him in Loritja, he asked about Lasseter. No, he didn't know him.

'Lion then said to him: "This man [Buck] is not a policeman. This white fellow has got a station and everybody is working for him. He's not a policeman. We want to find that other man [Lasseter]."

'Anyhow, that old man say: "He died that way."

'Lion then came over that night and told me. In the morning I informed Buck: "Hey, that boy bin tellum that man Lasseter is dead out here."'

'Oh, that's all right,' said Buck. 'Now we know where to go.'

Buttons also recalled the search party heading out to locate the body: 'This old man . . . he now acted as a guide and showed us the way . . . we unloaded all our goods and camel gear, everything and hobbled the camels. We took only five with us and went on to Lasseter's grave. We followed up the creek and got there at dinner time. It's only half a day from there to where he died.

'The old man said: "He's there. He's finished." We asked him to show us the grave, so he came over and showed us, "That's the grave under the tree."

'He was afraid to come any closer . . . The local people who buried him only dug a shallow hole; when they laid him on his side, the body was all doubled up. Then they threw a bit of dirt on top and later a few branches.

'We threw all these away and pulled him out. It was bad. Every time we pulled, something would give way [break]. We shovelled everything out and dug a deeper hole alongside. We found maps, letters to his wife, his false teeth, and a tomato sauce bottle with a letter in it buried under a tree on a sandhill. Then we reburied him. We didn't put a fence around the grave, just put in one post.'

In a pencil-written letter Buck described how he had buried the decomposed body together with an arm 'the dogs had taken'. He told of how the local Aborigines said that Lasseter had died of starvation. There were no signs of violence.

•

It seemed John Bailey's prayers had been answered. Lasseter's death would give the company chairman the keys to the vault at the Bank of Australasia in Sydney and possession of the closely guarded coordinates to the reef's precise location, written in invisible ink in the prospector's own hand. Lasseter was irrelevant. Now the union boss could play his ace, and he wasted no time.

But Bailey and his son Ernest were soon to discover that like everything else that had taken place during this never-ending hunt for gold, obtaining Lasseter's mysterious document would not run smoothly.

As Bailey wrote in his account 'The History of Lasseter's Reef', 'When the Directors were satisfied that Lasseter had perished, the Chairman of the Company and the Secretary of the Company . . . arranged an interview with the Bank Manager to see if he would hand over the particulars to the company.' But to their astonishment – and frustration – the bank manager, like many Australians, had the vague suspicion that Lasseter might actually be alive somewhere. 'The Bank Manager said: "Are you sure he is dead? If so I want to see the death certificate from some qualified person."' They produced the burial certificate as signed by Adolf Heinrich at Hermannsburg, but the bank manager declined it as insufficient proof, saying, 'I refuse to hand the documents over.'

Bailey and son were stumped. No stranger to thinking on his feet, the union boss replied that the reason he needed to ascertain Lasseter's coordinates so urgently was to 'find him before he was dead if possible'. It cut no ice. There wasn't an option – the company was 'compelled then to make an application to the Supreme Court for the recovery of this document'. In time, Lasseter's letters were released to the public trustee, whereupon 'in the presence of interested parties it was opened'.

A Detective Sergeant Thompson from the NSW Criminal Investigation Branch was brought in to cast his eye over the document to see if it did indeed contain a message written in invisible ink. Sergeant Thompson concluded it did: the document could have been written with 'uric acid, milk, lemon water or limewater'. He 'gave directions to treatment, and as a result a second set of directions was brought to light'. Yet these proved to be worthless – apparently a set of compass bearings and variations, but with no reference points given to which they could be applied, the message might as well have been a mishmash of random numbers.

Even after he was dead Lasseter was making Bailey's life hell – but one way or another the union boss was going to make the prospector pay. The company had taken out an insurance policy on Lasseter for £500 and now that he was dead, Bailey decided to cash it in. But like the Bank of Australasia, the insurance company wanted proof that the prospector was indeed deceased. What they needed was a statutory declaration from the man who had found and buried the body out at Irving Creek, Bob Buck. But when the directors approached the old cattleman they were in for a shock. According to Bailey, when Buck was asked to make the statement he 'refused to do this. He said he could not swear whether the skeleton was that of a white or black man.'

Bailey was stunned. Buck's admission that he wasn't even sure of the race of the body he'd buried was not something anyone was expecting.

To compound matters, upon notification of Lasseter's burial, the CAGE company had ceased the stipend payments made to the prospector for his services as the expedition guide. In his 'History of Lasseter's Reef', Bailey recounts that 'the alleged Mrs Lasseter became very annoyed . . . and made a statement to the *Truth* newspaper'. Bailey was visited by a reporter who asked him to respond. Bailey recalled, 'I asked him if he were sure he had been talking to Mrs Lasseter or the alleged Mrs Lasseter . . . I thought he would be well advised to call on his informant and ask her to show him her marriage certificate.' Bailey was pleased with the outcome: 'He then wrote the article, "Is this woman Mrs. Lasseter?" From then on we received no further request from that quarter.'

33

STRANGE TURNS

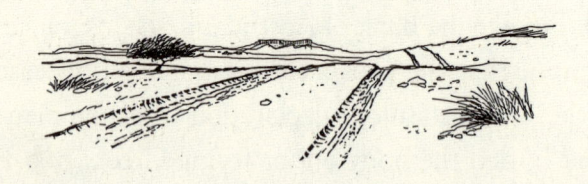

There was a final roll of the dice for the Central Australian Gold Exploration Company. A new mission to find Lasseter's reef was set up, instigated and financed by a business entrepreneur, Leslie Bridge, who intended to film the operation as a documentary. Bridge's motivations were genuine. In 'the national interest' he approached John Bailey to reinvigorate the search, into which he poured thousands of pounds of his own money. What became known as the Second Expedition would be led by the now nationally celebrated Bob Buck, with six or seven other expeditionaries, including a cinematographer. This time there would be no troublesome trucks or aircraft. Camels would be the order of the day. The team set off from Alice Springs in September 1931 and proceeded to Ayers Rock and Mount Olga, eventually reaching the Pultardi Rockhole around a month later.

Having set up camp near a cave on the Hull River on the northern edge of the Petermann Ranges, Buck suddenly spotted an inscription carved into the bark of a tree outside a cave. It read, 'Dig Floor Top End Attacked'.

Buried beneath the remains of a fire within the cave, Buck discovered something extraordinary: a sealed notebook that was more or less a diary detailing Lasseter's last days. It was a simple red fabric-covered pocketbook measuring about five by six inches. The elements, the soil and all-pervading white ants had not been especially kind to the document, but the book's sensational contents managed to fill in many of the blanks as to what had happened to Harold Bell Lasseter.

News of Buck's remarkable discovery was transmitted back to Sydney using the expedition's fully functioning transmitter, and on the mission's return to Alice Springs the precious book was immediately posted to head office. But apart from unearthing Lasseter's diary, a broken revolver and other sundry items, this Second Expedition had been nothing short of a fiasco, and it was the last CAGE foray into the dead heart, the enterprise supposedly breaking down because of the quantity of alcohol taken in place of decent food supplies.

But as for Lasseter's diary, it is a remarkable record. It is not exactly a logged daily journal, but rather a collection of observations and reflections, simple maps and drawings created in pencil. The old prospector dramatically described his worsening predicament, subsisting among the natives, his fear of being murdered, his longing to see his wife and children back in Sydney, his anger at Fred Blakeley, his frustration with Paul Johns, his deteriorating health – trachoma and malnutrition – his terrible hunger and his despair at being left to die. The diary is certainly compelling, and through it, Lasseter's last, lone movements can be pieced together. Many of the sentences are illegible or incomplete due to the book's deterioration but his story nonetheless comes through.

It begins dramatically with his description of how 20 Aborigines 'woke me up, and told me they were going to kill me, the headman who is so treacherous being the leader and spokesman . . . I expect an attack in force tonight . . . In the dust of the cave I have just unearthed an unexpected find – 5 revolver cartridges, tho' as they know I am practically blind that may not avail me much, two or more spears were carried by each man.'

It is indeed a gripping and harrowing read. Lasseter's fear of abandonment is palpable through his handwriting, wondering what on earth has happened to Paul Johns as he wrestles with the idea of suicide. 'Why is no relief sent, what became of Paul. The suspense of not knowing is the worst of all, why do I cling to life when a shot would end my torment. It's just because I want to know why everyone has failed me. To die a lonely horrible death is bad, but not to know why is worse. It is now 25 days since the camels bolted.

'Of course I am myself to blame for tackling this job alone, but I thought I could trust the blacks not to do anything worse than raid me. Good Bye Rene darling wife of mine & don't grieve. Remember you must live for the children now dear, but it does seem cruel to die alone out here because I have been too good to the blacks my last prayer is "God be merciful to me, as sinner & be good to those I leave behind. XXX Harry X"

'How I long to see my children once more, to hold their chubby hands and to see their laughing faces and hear their baby prattle. My God why does not help come, with lots of water I can hold out for several days yet but the agony of starvation may drive me to shoot myself . . . I should never had gone on alone but I relied on Paul to follow me. What good a reef worth millions. I would give it all for a loaf of bread . . . I have just measured my chest, I have shrunk from 39 to 25 inches and my waistline has an even greater shrinkage.'

Occasionally Lasseter would return to his diary, with his eyesight failing, again to write to his wife. 'Dear Rene, I think I am near my finish. I am nearly . . . and crazy with sandy blight . . . tormented with flies and ants . . . I have done my best darling and am sure I could have got out of it had not been . . . this sandy blight.'

As his condition deteriorated further, an Aboriginal elder showed him compassion: 'Later an old chap with a wart 6 inches by 3 inches in the fold of his posteria [sic] took pity on me & brought 2 rabbits & some berries like cape gooseberries, my eyes are still blurry & smart a good deal but are much better than last week.'

Then, within the fragile pages, Lasseter makes some startling disclosures: 'Darling I've pegged the reef and marked the exact locality on the map which is buried in my kit . . . on the sandhills where the camels bolted – on the East side of the hill, and I photographed the datum peg dated 23rd December. I can't understand . . . support or relief has not been afforded me . . . I buried 3 rolls of film in a 5 lb treacle tin on the sandhills too.'

With no sign of any relief, he makes his plan: 'I'm now going to try the 80 mile dash to Mt. Olga, will travel all night . . . Give my love to the children and may we meet again. Darling I have always believed in a God, a supreme ruler of the Universe, but I have gone my own way so long that I'm ashamed to pray to him now . . . I'm just a skeleton now . . . now they are waiting for me to die in order to steal my shirt and trousers off the body.

'Of course I was a fool to take this on alone but I relied on Paul Johns overtaking me in 4 to 6 weeks at the outside. He averred that he would overtake me in three weeks and gave me his word of honour not to let me . . . Also it was agreed upon . . . Fred Blakeley when I engaged to go with the camels that if I did not show up again by the end of November that the . . . would send a man named Johansen to my relief. As I believe he also stumbled onto this identical reef. I had to go right out to Lake Christopher which is 100 miles across the W.A. border to get my bearings then I was . . . to go direct to the reef . . .

'I've tried to make you happy and if the Company treat you right you will be rich. The reef is a bonanza and to think that if only Fred Blakeley had been guided by me we could have got through with the truck in three . . .

'This is cruel to die of starvation, heartless of all who know I am out here. When I didn't return by Xmas they knew there was something wrong. May God forgive them and strengthen me in my last hour. No food for 2 days . . .

'Rene darling. Don't grieve for me I've done my best & have pegged the reef, not strictly according to law as the blacks pinched my miners right & I don't know the number but I photographed the datum post on the Quartz Blow. The post is sticking in a water hole & the photo faces north. I made the run in 5 days but the blacks have a sacred place nearby & will pull the peg up for sure. I've taken the films & will plant them at Wintersglen if I can get there the Blight has got me beat . . .

'Darling I want you to remember me as when we first met & not the scarecrow that I now am . . . have shrunk still further & the flies & ants have nearly eaten my face . . .

'Take good care of Bobby, Betty & Joy please I want Bobby to be a Civil Engineer try & educate him for that. Darling I do love you so I'm sorry I can't be with you at the last but God's will be done. Yours ever x Harry xxx'.

•

In shocking, lurid detail, the painful narrative of the prospector's protracted death was revealed to an appalled nation. Yet for some, his tragic end opened up the possibility of new beginnings.

Lasseter's exhumed diary stated that he had actually found and pegged his long-lost reef. Further, he had travelled all the way out to Lake Christopher across the Western Australian border; and again, there was a mention of this fellow Johansen – the same name in the letter sent with Paul Johns to the Government Resident.

From information gleaned from the letters found on Lasseter's body and the notebook diary, and much later from Aboriginal eyewitnesses, Lasseter's movements after Johns left him at Ilpbilla were pieced together, telling of the prospector's miserable demise.

It seems Lasseter remained at the old airstrip for perhaps a week. Local Aborigines were wary of him. Either to deter or impress them, he used discarded fuel drums for target practice with his pistol. It is likely he followed his old camel tracks to the Mount Bowley area, where there was a known water supply, and then pushed westward along the northern edge of the Petermann Ranges, passing the Hull River, and the cave in which the diary was later found. Eventually Lasseter and his two camels crossed the Western Australian border to reach Lake Christopher, the designated rendezvous point where he was to meet the mysterious Johansen. For whatever reason, the meeting did not take place, and Lasseter turned back along his tracks, heading north-east towards the Hull River. After passing the cave again, he continued to head east for about 10 miles, when on top of a sandhill he was struck with dysentery. While relieving himself, Lasseter left his camels unhobbled and with their reins free, only about 30 feet away. When he stood up, the camels spooked and bolted through the scrub, leaving him literally with his pants down. In desperation, Lasseter ran after them firing his revolver, hoping to shoot one to recover some of his supplies. He watched in horror as the camels loped into the distance, taking with them everything he owned.

Much credit should be given to author Billy Marshall-Stoneking, who had the foresight to record interviews with Aboriginal elders while living among the Pintupi people at Papunya Settlement in the Northern Territory in the early 1980s. His interview with Leslie Tjapanangka reveals that the camels were possibly 'spooked by two Aboriginal hunters who had stood some two hundred metres away watching the drama run its course, but in another version of the story,' writes Marshall-Stoneking, 'the camels were said to be scared by the noise of Lasseter relieving his bowels. At any rate the two hunters were members of a larger group camped nearby and they returned to tell the story of what they had seen to the others.'

Lasseter discovered not all was lost. Following the camels' tracks, he found that some of his equipment had been thrown by one of the animals over a distance of several hundred yards: some tea, sugar, Oxo tins, a 100-pound bag of flour, diaries and letters, his camera and a bag of gold samples.

Leslie Tjapanangka recalled how a group of Aboriginal men found Lasseter sitting on the ground among what few items he had salvaged, and taking pity on him, brought him back to the cave on the banks of the bone-dry Hull River. The cave, for the moment, gave Lasseter much-needed shelter, but in time it would seemingly become his prison, as day after day his chances of being rescued diminished. He lived among the local Aboriginal community, waiting for the sound of an aircraft or an approaching car, but nothing arrived.

Stricken with sandy blight and suffering from malnutrition and dysentery, he gambled all to undertake the 80-mile trek to Mount Olga. Thirty-five miles from the cave on the Hull River, Lasseter died.

In his book *The Fortune Hunters*, Frank Clune relates how he spoke to an Aborigine named Mickey, who was 20 years old at the time of Lasseter's demise. Mickey was with the prospector through his last moments. 'He all the time open book and write 'im. He sit along creek and drink plenty water . . . By and by that old man Lasseter he die. There was big mob of men, women, girls, boys and babies. We put 'im in ground, face looking upwards and his feet pointing west. He not naked, he wear 'im shirt and trouser. He no wear hat, he lose 'im long time ago. Him wear 'im boot. We all cried . . . and cover him with ashes and dirt.' It is thought the date was 31 January 1931.

•

It is generally accepted that Lasseter's fate played out in a fashion something along these lines. But the diary's accounts of 'treacherous' and 'hostile' Aborigines don't correlate with the Aboriginal memory of Lasseter. According to Marshall-Stoneking, 'Several Aboriginal people who as children had camped near Lasseter at the cave remembered him as a good natured man who was very good to all the children, making damper for them and drawing pictures'.

The diary seems to neatly cover off everything Lasseter wanted to say: his suffering, his love for his family, his remorse, his 'get-squares' with those he disliked, where he had travelled; and the greatest I-told-you-so of all, that his reef really did exist – 'to think that if only Fred Blakeley had been guided by me we could have got through'.

In accounting for Lasseter's final weeks the diary and its contents seemed too good to be true; and there were some who were convinced it actually was. Its sudden appearance was a surprise to Fred Blakeley, who had been with Lasseter for the entirety of the expedition until he was handed over to Paul Johns, and insisted he had never seen the red notebook before. He only knew of the big diary in which Lasseter would nightly pen his thoughts within the cab of the Thornycroft truck, as he had described: 'a nice-looking volume about eight inches long, six inches wide and perhaps two inches thick. It was well bound, with a very fine black morocco cover and exceptionally good paper.' Other questions arose as to how clean the diary's paper appeared. Even though there was marked deterioration, the book didn't seem to have the sweat-stained provenance of a document that for weeks had been scribed in and handled in a cave on the drought-stricken Hull River. And the handwriting seemed remarkably composed for someone writing how 'the flies & ants have nearly eaten my face'.

The letters Buck had initially recovered from Lasseter's body certainly seemed suspicious to Blakeley, who happened to be in the Sydney office when they arrived by post. 'He [Buck] sent in a few pages written in lead pencil saying that he found these on the body of Lasseter . . . There were only three or four pages, telling how Lasseter's camels had bolted . . . I queried the handwriting and the paper and suggested it be submitted to an expert. I thought that those pages came from a very cheap notebook, something like what kiddies buy for a penny. All the paper we took on the expedition was very good, and I was also sure that these pages did not come from Lasseter's diary, which was composed of very good quality paper.'

The red diary and the letters were not the only questionable aspects shrouding Lasseter's demise. For some, there was an uneasy feeling that Bob Buck's discovery of Lasseter's body was perhaps just that bit too convenient. It was said that Buck could be found propping up the bar of the Stuart Arms Hotel, 'regaling any passing tourist who would buy him a few drinks with his story of finding Lasseter's body'. But when pressed about whose body was actually in the grave, his storytelling went to water. Buck's reluctance to sign a statutory declaration about Lasseter's death raised as many questions as it did eyebrows. Even though the police in Alice Springs seemed satisfied enough with Buck's verbal claim, Walter Gill, who had travelled with Buck out to Lasseter's grave in May 1931, wasn't so sure. After Buck gave his spoken statement, Gill decided to corner the old cattleman about what really happened. He followed Buck to his room in the Stuart Arms Hotel, where 'in desperation, again I tackled Buck, who was now content to lie back on the bed eyeing vacantly a ribbon of cigarette smoke on its leisurely journey to the ceiling. I asked, did he sign a document? Had he made a written statement? . . . However each time a death certificate was mentioned, he dived for cover. It was something for one reason or another he wished to forget.'

Blakeley, for one, did not believe for one minute that Buck had found Lasseter's body. 'I was having a day out in Sydney some time later and having a drink in a hotel when in walked Buck. I asked him to have a drink, and said, "I know you well enough but can't find a name for you."

'He said, "Why, I'm Bob Buck from Central Australia." Now everything was clear. He only wanted to talk about cows and bullocks but I said to him, "Look here old man, you never found Lasseter's body. Harry was too well equipped to go out the way those diary leaves said, and why was it you did not find the two camels and the loading?"

'He said, "God only knows where they are."

'I said, "You know as well as I do that a bolting camel would throw its loading within three hundred yards. That's a very thin yarn to say you could not find them."

'He asked me who I was, and I told him. He was greatly upset for he had many warnings not to talk to me. "You won't crab a fellow's pitch, will you?" he said.

'I was pretty disgusted and left him,' Blakeley wrote, but Buck followed him into the street, pleading that he had lost so much stock during the drought and that at last he was turning a quid. 'If I can hang onto this game for a few more weeks I'll be able to restock when it rains.'

Blakeley said he had seriously considered heading straight for the newspaper office, but in the interests of Lasseter's wife and children chose not to.

In his book *The Search for Harold Lasseter*, Murray Hubbard reproduces a letter written in 1960 by Lasseter's daughter Ruby – from his first, and only legal, marriage – to her half-brother Robert. In it she mentions meeting Bob Buck just after he'd found their father's body, when he said that 'after he had seen the crowd in Sydney who had financed the expedition – his impression . . . is they did not want him found alive as there was a lot of skulduggery had been going on'. Certainly, as far as John Bailey was concerned, Lasseter's body had been considered the key to unlocking the vault in the Bank of Australasia. And if a body was needed, one wouldn't be hard to find.

Former ABC documentary maker Mark Hamlyn recalls his time in Alice Springs during the early 1980s, researching for a television series about the history of gold mining in Australia. While he was interviewing some of the Alice's old-time prospectors, then in their seventies and eighties, the conversation eventually turned to the discovery of Lasseter's body by Buck. 'The general tone of the conversation was that in those days anything could happen out there in the desert,' Hamlyn recalled, 'and that you could produce a skeleton out of pretty much anywhere. These old-timers were saying, ". . . well, you know what old Bob Buck was like – if he needed a body, he could have just shot a blackfella."'

That Buck was capable of murder to fill Lasseter's shallow grave was a serious accusation, and testaments from those who

Fred Colson's Chevrolet truck, carrying fuel for the Thornycroft which carried fuel for the plane. All the vehicles suffered terribly – eventually the truck suffering a broken pinion, putting the expedition's chances of success in jeopardy. (*Mitchell Library, State Library of NSW*)

The wreckage of the pretty *Golden Quest* after Errol Coote's spectacular crash at Aiai Creek. That Coote was not killed was nothing short of a miracle. (*Mitchell Library, State Library of NSW*)

Expedition leader Fred Blakeley's skepticism as to the worth of aircraft seemed more than vindicated. To his astonishment, another would soon take its place. (*Mitchell Library, State Library of NSW*)

Fred Colson's Chevrolet truck en route to Alice Springs, bringing what was left of the *Golden Quest*. Colson had to rescue Coote from the wreckage with an axe. (*Mitchell Library, State Library of NSW*)

Expedition members were uneasy with the sudden appearance of a mysterious young German dingo scalper named Paul Johns and his two Aboriginal guides. Even though Fred Blakeley was suspicious of Johns, he ultimately allowed Lasseter to travel alone with him.
(*Mitchell Library, State Library of NSW*)

Pilot Errol Coote's disastrous decision to relocate the expedition's base camp at Ayers Rock almost cost him his life. It was a pointless and costly exercise that achieved nothing except to delay the search for Lasseter further.
(*Mitchell Library, State Library of NSW*)

After expedition mechanic Phillip Taylor had repaired the *Golden Quest II*'s propeller, Errol Coote took the aircraft for a test flight before heading back to a hostile reception at Alice Springs. (*Mitchell Library, State Library of NSW*)

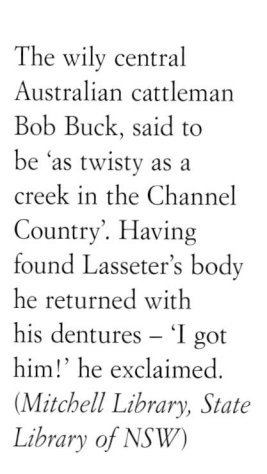

The wily central Australian cattleman Bob Buck, said to be 'as twisty as a creek in the Channel Country'. Having found Lasseter's body he returned with his dentures – 'I got him!' he exclaimed. (*Mitchell Library, State Library of NSW*)

Buck's discovery of Harold Lasseter's body turned him into an overnight celebrity. He was the city person's ideal of an Australian bushman. (*Mitchell Library, State Library of NSW*)

Buck noticed this carving on a tree near a cave on the Hull River. It led to the discovery of Lasseter's diary. (*Mitchell Library, State Library of NSW*)

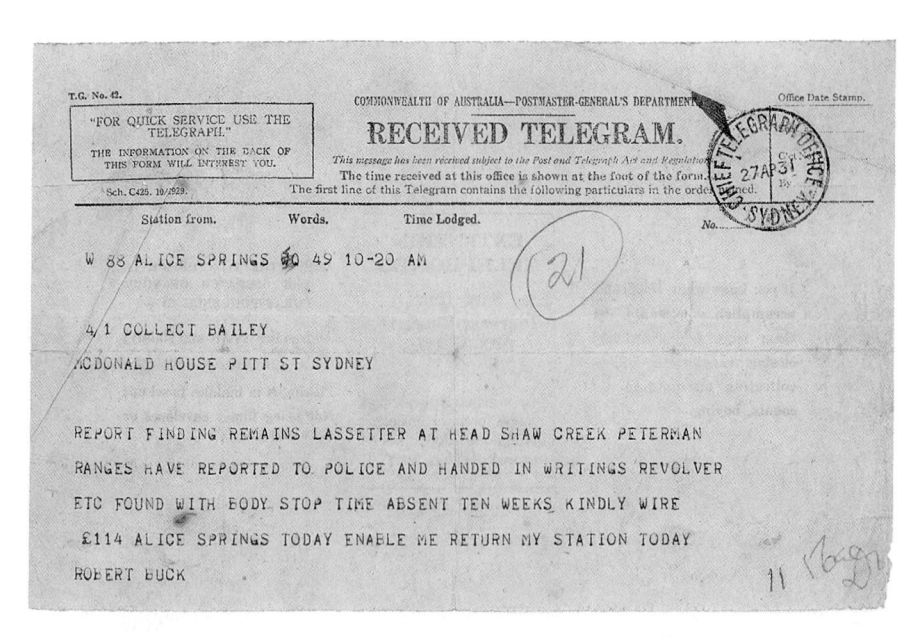

```
T.G. No. 42.                    COMMONWEALTH OF AUSTRALIA—POSTMASTER-GENERAL'S DEPARTMENT          Office Date Stamp.
"FOR QUICK SERVICE USE THE
        TELEGRAPH."                    RECEIVED TELEGRAM.
THE INFORMATION ON THE BACK OF      This message has been received subject to the Post and Telegraph Act and Regulations
    THIS FORM WILL INTEREST YOU.      The time received at this office is shown at the foot of the form.     27AP31
Sch. C425. 10/1929.                   The first line of this Telegram contains the following particulars in the order named.

            Station from.        Words.        Time Lodged.                              No.

    W 88 ALICE SPRINGS  40  49  10-20 AM                           (21)

    4/1 COLLECT BAILEY
    MCDONALD HOUSE PITT ST SYDNEY

    REPORT FINDING REMAINS LASSETER AT HEAD SHAW CREEK PETERMAN
    RANGES HAVE REPORTED TO POLICE AND HANDED IN WRITINGS REVOLVER
    ETC FOUND WITH BODY STOP TIME ABSENT TEN WEEKS KINDLY WIRE
    £114 ALICE SPRINGS TODAY ENABLE ME RETURN MY STATION TODAY
    ROBERT BUCK
```

Not seen for decades, Bob Buck's telegram to John Bailey announcing his grisly discovery. (*Courtesy of the estate of R & J Stalley*)

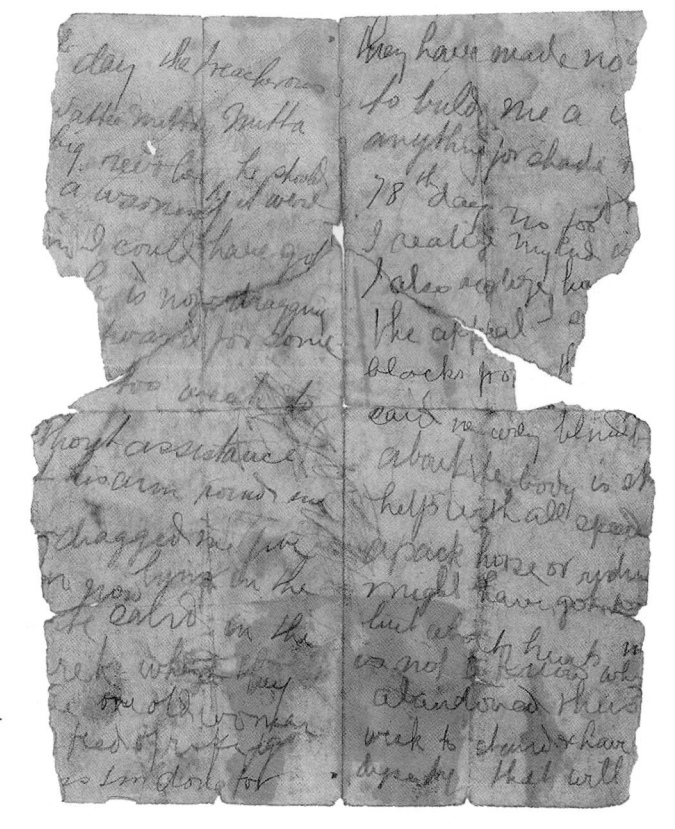

The letter supposedly written by Lasseter and discovered by Bob Buck, recently found in the estate of Richard and Jill Stalley. (*Courtesy of the estate of R & J Stalley*)

The news of Lasseter's death gave John Bailey the key to the vault of the Bank of Australasia and possession of the secret letter written in invisible ink giving the reef's location. (*Author's collection*)

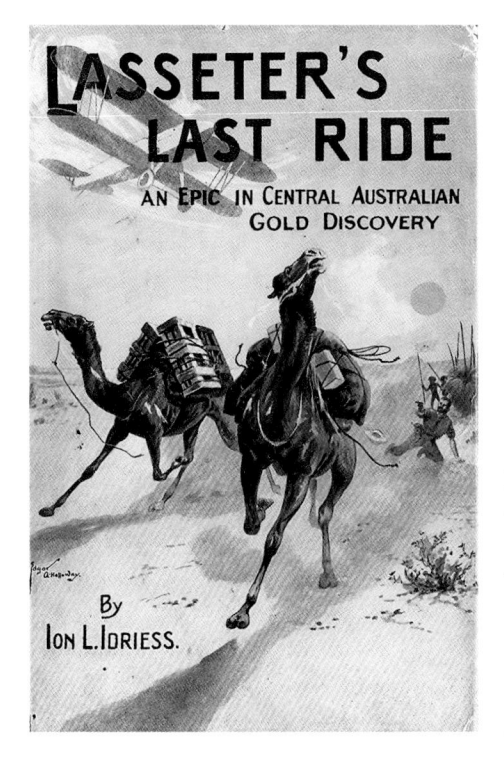

Ion Idriess's *Lasseter's Last Ride* was published barely six months after the prospector's demise and became an instant bestseller. The cover illustration shows much *Boys' Own* action – the book was largely responsible for fuelling the Lasseter myth. (*Author's collection*)

Queensland police photos of Paul Johns, the young German dogger was the last white man to see Harold Lasseter alive. Violent and a petty criminal, he was ultimately interned in Britain as a Nazi saboteur. Did he leave Lasseter in the desert to die? (*National Archives of Australia*)

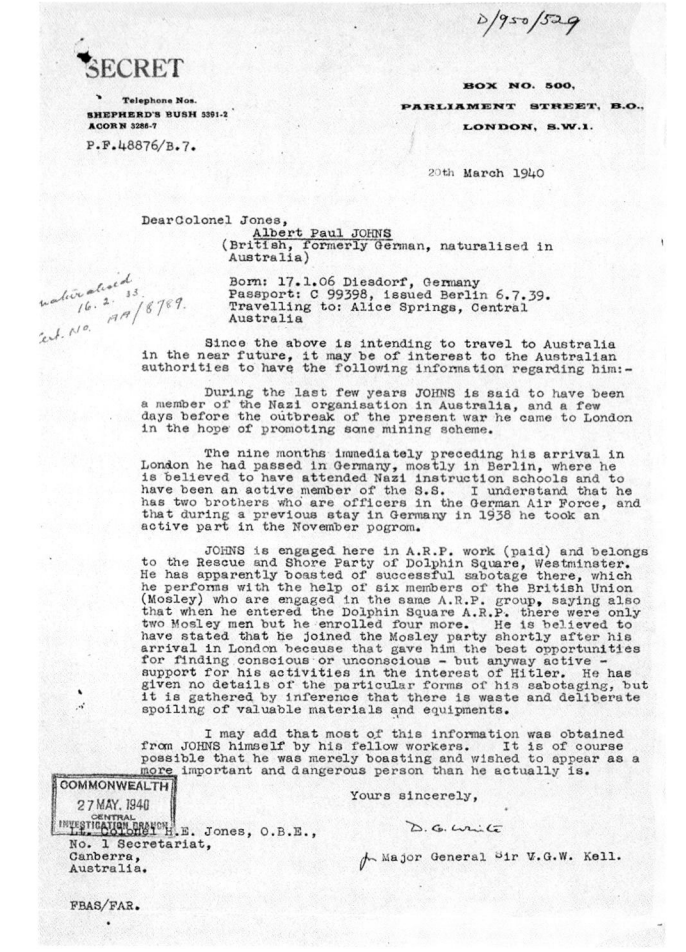

SECRET

Telephone Nos.
SHEPHERD'S BUSH 3391-2
ACORN 3285-7

P.F.48876/B.7.

BOX NO. 500,
PARLIAMENT STREET, B.O.,
LONDON, S.W.1.

20th March 1940

Dear Colonel Jones,

Albert Paul JOHNS
(British, formerly German, naturalised in Australia)

Born: 17.1.06 Diesdorf, Germany
Passport: C 99398, issued Berlin 6.7.39.
Travelling to: Alice Springs, Central Australia

Since the above is intending to travel to Australia in the near future, it may be of interest to the Australian authorities to have the following information regarding him:—

During the last few years JOHNS is said to have been a member of the Nazi organisation in Australia, and a few days before the outbreak of the present war he came to London in the hope of promoting some mining scheme.

The nine months immediately preceding his arrival in London he had passed in Germany, mostly in Berlin, where he is believed to have attended Nazi instruction schools and to have been an active member of the S.S. I understand that he has two brothers who are officers in the German Air Force, and that during a previous stay in Germany in 1938 he took an active part in the November pogrom.

JOHNS is engaged here in A.R.P. work (paid) and belongs to the Rescue and Shore Party of Dolphin Square, Westminster. He has apparently boasted of successful sabotage there, which he performs with the help of six members of the British Union (Mosley) who are engaged in the same A.R.P. group, saying also that when he entered the Dolphin Square A.R.P. there were only two Mosley men but he enrolled four more. He is believed to have stated that he joined the Mosley party shortly after his arrival in London because that gave him the best opportunities for finding conscious or unconscious – but anyway active – support for his activities in the interest of Hitler. He has given no details of the particular forms of his sabotaging, but it is gathered by inference that there is waste and deliberate spoiling of valuable materials and equipments.

I may add that most of this information was obtained from JOHNS himself by his fellow workers. It is of course possible that he was merely boasting and wished to appear as a more important and dangerous person than he actually is.

Yours sincerely,

D. G. White

Major General Sir V.G.W. Kell.

Lt. Colonel H.E. Jones, O.B.E.,
No. 1 Secretariat,
Canberra,
Australia.

FBAS/FAR.

Letter from Major General Vernon Kell to the No. 1 Secretariat in Canberra regarding Johns. (*National Archives of Australia*) were with him at the time of the body's discovery discount it. But in some quarters, it was not considered beyond the realms of possibility. Life was cheap out in the dead heart in 1930 – particularly if you were Aboriginal.

•

The Lasseter story was taking some strange turns. The notion that the body in the grave at Irving Creek could be that of someone else was certainly a bizarre twist. But this gave traction to an even stranger possibility: what if Lasseter was in fact still alive? For Blakeley, a lot of the ill-fitting pieces of the puzzle were starting to fall into place. While he was satisfied that Lasseter wrote the contents of the questionable red diary found by Buck, Blakeley was convinced that someone had deliberately planted the manuscript so that it could be conveniently found. As far as he was concerned, Lasseter had been up to something. In 1955, Blakeley, who had spent the past 25 years collecting whatever information he could about what had happened out in the desert all those years ago, broke his silence in a magazine interview. In short, he believed Lasseter was alive and living in the United States.

Blakeley 'claims Lasseter was never a bushman, knew nothing of mining or mineralogy and never had a gold reef', reads the article. 'Even more provocatively, he claims that Lasseter did not die on his lone ride to locate the reef after the rest of the expedition had turned back ... He says that Lasseter rode safely out of the desert, crossed to Western Australia and got away to America.' It was certainly a big claim, but Blakeley didn't stop there. He was convinced that Lasseter had 'flitted' and was well and truly alive and kicking, having linked up with the Mormon Church in the United States. He claimed, 'Mormons in Australia told him for some years he [Lasseter] had been a Pastor in Salt Lake City, Utah.'

But the question was: why would Lasseter orchestrate such an elaborate escape? Blakeley said he believed that Lasseter 'wanted people to think he was dead because he knew he would never be able to produce his reef . . . if he had gone back he might have been denounced as a mountebank'. The interviewer inquired, 'If Lasseter's story were false, the question is, what were his motives in carrying out the fraud?' According to the article, both Blakeley and Phillip Taylor – also questioned – agreed: 'They say Lasseter was in financial straits and pitched a tale to obtain money (the company paid him £10 a week during the expedition). They believe that Lasseter hoped that once in the Centre, he might accidentally stumble on gold-bearing country.'

Blakeley had worked out Lasseter's escape plan: 'he got on to the well-beaten stock-route from Hermannsburg to Eucla and followed it into Western Australia.' Apparently Blakeley had contacted friends in Eucla and given them Lasseter's description, asking if anyone had passed through who fitted the bill. The answer was yes. 'A man answering that description had come in from the north and continued on to the west.'

Others supported Blakeley's conspiracy theory that Lasseter wasn't dead but actually on the run. Reports began to surface from all sorts of people claiming they had seen him too. In 1931, a man named Tyler from Windimurra Station in Western Australia was certain he had seen Lasseter on the Canning Stock Route. To support Tyler's story, this mysterious fellow told him he was a surveyor. A letter written to Blakeley held in the Mitchell Library tells of a mining engineer who had heard Lasseter speak at a CAGE meeting in Sydney before the expedition was formalised. Sometime after Lasseter's death, the engineer was visiting Oregon in the United States and was shocked to see Lasseter standing on a wharf checking timber. Upon speaking to him, the engineer 'was informed by the man . . . [that he] was the half-brother of Lasseter'. Another company shareholder said she had also seen him in the US, travelling aboard a steamer 'under the name of Bell'.

There are some stories that are so compelling, they are hard to ignore. In the Mitchell Library is a series of letters addressed to Fred Blakeley, written in 1956 by Mrs Nellie Edwards of Mullewa in Western Australia. She revealed a remarkable story, explaining that she had seen a picture of Lasseter accompanying an article about the gold reef in an old magazine. The photograph greatly upset her. She was sure it was a picture of a man named 'Duncan' she and her husband had met at her mother's boarding house in Wiluna, Western Australia, during 1931.

During a conversation one evening, Mrs Edwards's husband, a gold miner, showed Duncan a few gold samples, upon which the visitor went off to his room and returned with 'two really rich specimens'. Duncan explained that these gold samples had come from Lasseter's reef. He had been with a party of men who had stumbled onto the prospector's body, and these samples had come from a bag found alongside it. But he said he had not seen the reef himself.

That night Duncan brought a brown paper bag filled with scraps of paper into the boarding house kitchen and asked Mrs Edwards's mother if it could be burned in the stove. It was just a few old letters and some paperwork he wanted to get rid of. When he left, the old lady stoked the fire, prodding the bag as it burned in the hearth. Mrs Edwards noticed a wedge-shaped remnant of a torn envelope that had fallen to the floor. 'I picked it up,' she wrote. 'One complete word was on it: "Lasseter".'

There were other strange and intriguing occurrences during the boarder's stay. Mrs Edwards picked up a small photograph that had fallen out of the visitor's wallet, depicting him standing with a woman and two small children. She asked him if they were his family, but he said they were just 'good friends'. 'He had tears in his eyes,' she remembered.

What Blakeley, or any of the others, did not know was that Lasseter had faked his death before – a fact not known until the *Australian* newspaper revealed it in April 1989. The opening paragraph read: 'Harold Lasseter – whose so-called Lost Reef has long captivated the imagination of Australia – was a confidence trickster, compulsive liar and probably a bigamist who faked his death during WWI to escape an unhappy marriage.'

The article claimed that Lasseter had 'indulged in an elaborate hoax to convince his family he was dead'. The story was revealed by Mrs Alethea Wilson of the Sydney suburb of Earlwood, whose family in 1915 accepted a flat-broke Lasseter accompanied by his wife and daughters as boarders in their home. There was some sort of loose family connection by which the Lasseters were there, but 'he turned up on our doorstep out of the blue', said Alethea, a school student at the time; and 'they stayed and stayed and stayed'. While in Sydney, Lasseter was trying desperately to interest the army in his many schemes, particularly his plan to win the Dardanelles campaign, Alethea recalled – 'but the army saw no merit in it much to Lasseter's disgust'. Eventually his sister Lillian contacted him about the availability of a house in Adelaide, an offer that the Lasseters took up. But, as Alethea said, 'The family never reached Adelaide and dropped out of sight.'

It wasn't until November 1916 that the Wilson family heard anything more of Lasseter. A photograph of him in uniform appeared in the *Sydney Mail*, posted among those killed in action. The caption read: 'Sap. L. H. Lasseter, Meredith (Vic), died of wounds.' According to Alethea, Lasseter's sister Lillian tried to find out more about her brother's death, with no luck.

Years later, one of the Wilson daughters was working as a nurse in a maternity hospital, when in walked Lasseter carrying a bunch of flowers. 'He didn't recognise her, but she knew exactly who he was. She took the flowers from him, put them in a vase and took it to Mrs Lasseter. But it was not the Mrs Lasseter we knew, and now he was calling himself Harold Bell Lasseter.'

Perhaps the *Sydney Mail* merely made an error in publishing his photograph and details of how he died. But knowing Lasseter's penchant for changing his name, changing his wives and reinventing himself, it seems unlikely.

Certainly the obituary suggests that Lasseter had died serving overseas during the Great War, but his comprehensive military files in the National Archive show fairly brief and erratic army service and no indication he ever left Australia. Did he try to fabricate a scenario where he had fought on foreign battlefields? Curiously, a letter appeared in *The Bulletin* in 1932 from someone who knew Lasseter when he and his family lived at Tabulam. It is a particularly interesting letter for one casually mentioned detail: 'I haven't much faith in Lasseter's Reef,' it begins. 'I knew Lasseter. My first acquaintance with him was about 16 years ago when he was irrigation farming on a 10-acre lease at Tabulam, NSW. He claimed to be a Victorian by birth but had spent many years in the USA. He sold out his lease and enlisted. Writing to me from Gallipoli he told me he had invented . . .'

34

TORN APART

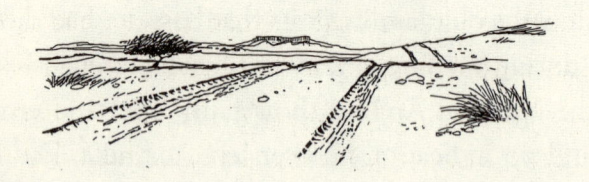

Amid all the frenzied speculation as to what had happened to the prospector, it was as if Harold Lasseter himself had been stripped of all dignity as a human being. It was as though he had existed merely as a facilitator for other people's wealth, and for many his disappearance and death was hardly a tragedy, rather another damned inconvenience. The expensive multi-aircraft RAAF searches that scoured Central Australia to rescue the inept flyers Pittendrigh and Hamre were never afforded to Lasseter. Paul Johns, who had left him in the desert, had had no intention of ever returning to help. John Bailey and his son Ernest had been on a crusade to prove Lasseter was dead so that they could get their hands on the letter in the bank. Bob Buck reburied someone he said was Lasseter out at Irving Creek, but he couldn't be certain whether the person was white or black. No-one seemed to consider that Harold Lasseter was once a living, breathing person and was survived by a living, breathing widow and living, breathing children.

The callous, roughshod tug-of-war surrounding Lasseter's memory plumbed new depths in 1957, when American documentary maker Lowell Thomas arrived in Alice Springs to begin filming for his television series *High Adventure*. Thomas was one of the most influential journalists of the twentieth century. He single-handedly introduced Lawrence of Arabia to the world in 1917, and became a television travelogue pioneer. Australia had only just begun broadcasting TV a year earlier, and Thomas was a big-name star by any standard. He had heard the story of Harold Lasseter and his reef, and that he might not in fact be dead but living in the United States. And so he decided he would find out firsthand if this was true by trying to locate the body – untouched since Bob Buck buried it in 1931. What was presented ostensibly as a kind of televised archaeological dig was in reality pure and simple grave-robbing.

Thomas and his crew are first depicted driving a convoy of Land Rovers meandering around the mulga until they reach the site of Lasseter's grave. With the help of two Aborigines – one of whom had been to the site some 26 years earlier – Thomas's team of outback blokes in flannelette shirts and bush hats locate the spot where Lasseter was said to be buried – and dig. They soon unearth a skeleton, and Thomas holds the yellowed skull aloft, Hamlet-like, and says to the camera in his mellifluous voice, 'Alas Harry, I didn't know you well – however there is one thing I do know. You may have been a failure, a fake and a fraud, but to have lived and wandered alone out here in this wilderness – you must have been a great bushman.'

The program, understandably, was apparently never screened in Australia. Lasseter's remains were placed in a bag and brought back to the police station in Alice Springs, where the program's producer Lee Robinson, production manager Curly Frazer and an Aborigine, Nosepeg Tjupurrula, were arrested for 'desecration of a grave'. In what was shaping up to be an international incident, the police were 'leaned on' by the powers above and begrudgingly released the miscreants.

Billy Marshall-Stoneking wrote about an interview he conducted with Nosepeg Tjupurrula, the Aborigine who had located the grave for the film. Marshall-Stoneking had heard the story of Lowell Thomas and the Lasseter exhumation many times, but on one occasion, when Nosepeg was describing parts of the body they unearthed, he mentioned Lasseter's teeth.

'But then I remembered that Lasseter didn't have teeth,' said Marshall-Stoneking. Bob Buck had brought back his dentures. Nosepeg was questioned about them: 'Are you sure he had teeth?'

'Yes, inside all the way,' replied Nosepeg, placing his fingers inside his mouth to show how the teeth were fixed. 'Inside all the way.'

'But Lasseter had false teeth.'

'Not false teeth! True one! Inside.'

Lasseter's freshly exhumed remains weren't always treated as so matter of fact. Author Thomas Keneally wrote of a conversation he had with the colourful old-time proprietor of the Alice Springs Hotel, Ly Underdown, then living in the Old Timers Home in Alice Springs in the early 1980s. Underdown said that when the body was brought back into town he 'registered the crated remains as a guest in the hotel and gave them a room'.

'Even so . . .' continues Keneally, 'Ly Underdown doubts that they were Lasseter's remains. "Remains of some bloody Abo, more likely. Lasseter had debts and a missus. Of course he flitted."' As Keneally comments, 'It's no use arguing that Lasseter could perhaps have flitted at somewhat less expense than taking an expedition to the Petermanns. Lasseter lives, O.K.'

•

But what of the mysterious 'Johansen, from Boulder City' whom Lasseter was supposed to meet at Lake Christopher, 90 miles across the border into Western Australia? On a camel expedition approaching the lake in November 1932, explorer Michael Terry and his party stumbled upon a tree in the Rawlinson Range with the name and date 'Lasseter – 2/12/30' carved into the trunk.

Then, while crossing the blinding salt expanse, Terry discovered a pole lying on the surface with a rag fixed to its top that had clearly once been planted in the crust vertically with three supports. Scribed in the salt below were the words 'DIG UNDER ---'. Terry wrote that he couldn't make out the last word – it was either FIRE or POLE. The expedition members righted the pole to reconstruct what it would have looked like and then dug underneath and around it, finding the remains of two campfires; they dug under those too, but found nothing. The place was a ghostly reminder of Lasseter – where he'd once been – now standing as a grim edifice. But, as Terry wrote, 'It establishes his presence with camels at Lake Christopher early in December.'

That Lasseter had actually made it all the way to Lake Christopher on his own beggared belief. As Errol Coote wrote, 'Paul Johns was emphatic that he would not be able to get through, as he did not have sufficient knowledge of the country. Too many people, including me, thought much the same – we were wrong.'

It seems, as the diary states, that Lasseter did indeed make his way alone to Lake Christopher, presumably for a rendezvous with the fabled 'Johansen', the name first surfacing in the letter Lasseter gave Johns to present to the Government Resident. The question as to who the elusive figure was has baffled historians, researchers and anyone remotely interested in the confounding Lasseter tale for the best part of a century. Detective work was befuddling. There were, of course, multiple spelling variations of the name – even 'Joe Hansen' could have been a possibility. And then there were the wild tales of misfortune. There was a convincing rumour in 1931 that two men – surnames Johansen and Fabian – were murdered by Aborigines somewhere near the Warburton and Jamieson Ranges, the *Perth Daily News* reporting that these two men from Alice Springs had been 'done to death' and eaten by wild natives. It was another mystery upon a mystery, but what emerged was that both John Bailey and Lasseter's wife Irene did in fact know of Johansen's existence before the expedition was underway.

It was not until 2010, when filmmaker Luke Walker was conducting research for his documentary *Lasseter's Bones*, that the missing piece of the puzzle materialised. Walker made a telephone call – a cold call – to Ann (Alvhild) Clark, whom he suspected could be related to a Swedish immigrant named Olof Johanson. As it turned out, she was his Australian-born daughter. Further, Alvhild's son Chris was a well-known Canberra-based defence historian who then took on the challenge of tracing his grandfather's past, eventually writing the book *Olof's Suitcase*. In tremendous detail, Clark delved into his grandfather's life, which in turn resulted in Perth researcher Mark Chambers's theory about how the relationship between Lasseter and Johanson came about.

When the CAGE expedition was announced to the press in May 1930, an article about the proposed gold hunt appeared in the *Perth Daily News*, naming Lasseter and his 'personal friend' John Bailey, 'President of the AWU'. It is believed that Johanson wrote to the union – as he was already a member – explaining that he also knew where the reef was. He'd seen it too. According to an old workmate of Johanson's, while hunting dingoes in the Rawlinson Ranges, Olof 'had accidentally stumbled upon a reef that Lasseter claimed to have discovered . . .' Apparently an

Aboriginal woman had led him down a spur and 'in this locale he came across a reef ... in which a shallow shaft had been sunk'.

There is documentation showing that Johansen then corresponded with Lasseter, who, as Clark writes, 'in all likelihood ... would have immediately shown the item to Bailey, possibly in order to quell any doubts his principal backer might have entertained about the veracity of his tale'. Certainly several letters exist, sent by Johansen to Lasseter. They are clearly in response to queries Lasseter had made about the area – Lake Amadeus, the Petermanns and Aborigines. In one, Johanson writes, 'No I have not known of a war party to number 1,000 not in my own experience but I have heard of such things talked about, some of it may be true ...' They certainly indicate that Lasseter had not been out there before. Due to the formal style of writing, it is very likely the pair never met, and certainly there is no mention of Johansen travelling out to Lake Christopher to meet Lasseter. But Clark suggests that in his desperation – and delusional state of mind – Lasseter may have become fixated on the idea of Johansen heading out to Lake Christopher to join the search. The pair of them would be able to find it. And so he sent the letter with Johns. But as we know ...

35

AWKWARD QUESTIONS

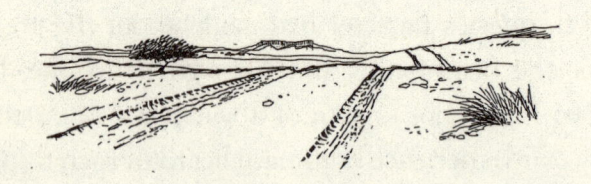

The tale of the grand, technologically advanced mining expedition that had spectacularly collapsed, leaving its chief protagonist dead somewhere west of Alice Springs, had seized the nation's imagination. Across the continent, Australians had all of a sudden become infatuated with the Lasseter story. It had all the ingredients of what we would now call an urban myth: a giant fable about the obsessive hunt for gold, a fabulous treasure ferociously held secret by cruelly disposed forces of nature – ancient, inscrutable and malevolent. And it told the tale of an old and eccentric prospector who had once found the amazing treasure, and who had perished trying to find it again.

Despite the story having all the hallmarks of a well-worn, far-fetched campfire yarn, as far as depression-gripped Australians were concerned, the gold reef was still bloody well out there somewhere, waiting to be rediscovered. Six months after the expedition was over, author Ion Idriess released a book detailing what had transpired with the ill-fated CAGE expedition. The book, titled *Lasseter's Last Ride*, was a bestseller in five editions, reprinted more than forty times since publication in 1931. It became something of an Australian bush classic and therefore was required reading in many schools. In an age when talking pictures were in their infancy, long before the advent of television, before *60 Minutes* or *Four Corners* could retell a story, *Lasseter's Last Ride* was the closest thing to a current affairs documentary. It was written as a kind of novelised nonfiction, and in the saga of the expedition traipsing backwards and forwards across Central Australia, readers were introduced to mysterious Aboriginal hoodoos, bone-pointing, imagined conversations between individuals, and – typically for its era – Aborigines sometimes wheeled in and out for high drama or light relief. Idriess imaginatively pieced together Lasseter's last months, based largely on the remnants of the pocketbook diary Bob Buck returned with. The hostile, guns-drawn confrontation between Lasseter and Johns is the book's climax, with the chapter titled 'The Fight in the Desert'. It's colourful stuff: '. . . Lasseter threw his plate of food at the other's face and they fought like fiends. One drew his revolver, but the other grabbed his wrist. They fought to their knees for that gun. The hammer jarred down between Lasseter's left finger and thumb. In pain he wrenched his wrist free and hurled the revolver into the spinifex. They sprang up, panting, mad-eyed, each awaiting each other's rush . . .'

Lasseter's Last Ride drew Paul Johns to the public's attention. Clara McKinnon, who knew Johns when he was a boarder at their family home at Clare in South Australia in the late 1920s, recalled, 'Friends of ours . . . suggested we buy the book *Lasseter's Last Ride*, because Paul Johns's name was in it, and they knew we had taken him into our home at one stage.' Idriess's book painted Bob Buck as the story's hero, while Johns – although not portrayed as a villain – certainly seemed a person of interest. Perhaps to defuse the gravity of 'The Fight in the Desert', when the book was at its selling peak in 1932, Johns gave a verbal 'statement' to prominent author and journalist Ernestine Hill, who described the interview as 'taken verbatim'. Certainly, the charismatic Johns had found a sympathetic ear in Hill. Historian Alan Powell in his history of the Northern Territory, *Far Country*, describes one of her works as 'wild romanticism, masquerading as history . . .' But Johns's statement suggests that he was holding his cards very close to his chest. At the time, he was in the process of trying to renounce his German citizenship to become a naturalised Australian, and the famous gunfight described in *Lasseter's Last Ride* was casting a shadow over his suitability, with police requests for information about his past not helping his case.

Ernestine Hill's publication itself is a curious document. It was published as what has been described as a 'pamphlet' in a relatively small print run, replete with stylised lino-cut illustrations, giving it the feel more of a children's book than a formal account from the last white man to ever see Lasseter alive. Hill prefaced Johns's statement with a fairly flowery introduction in which she described her German interviewee as 'then a tall and stalwart young Englishman of Nordic fairness . . . His story is briefly and simply told, as a bushman tells it, without drama or ordeal in that slow circle of a thousand miles, but with belated regret at leaving Lasseter alone. He could not do otherwise.'

Johns's statement is short – only five and a half pages – and gives a spartan account of where he and Lasseter travelled. He mentions they 'had a quarrel at Mount Stephenson. Watering the camels, I threw away what was left in the bucket. Lasseter shouted abuse at me, threatened me with a rifle. He said there could be trouble, the natives worshipped water in this country and our lives wouldn't be safe if they saw us throw it away. I said nothing, I let it blow over.'

To defuse the public's enthrallment with the fight allegation, Johns described Lasseter to Hill as 'generous and likeable, and a good mate'. Yet accounts from the expedition members – Blakeley, Coote, Taylor, Sutherland and Blakiston-Houston – make it clear that Lasseter was not generous, likeable or a good mate to anyone.

Johns told Hill that Lasseter's last words to him as he left were: 'Don't leave me alone for too long. Whatever passed between us is over now. If I wasn't sure of you, I wouldn't go on.' To which Johns replied, 'Look here Harry, I'll do my best. I'll be coming back as soon as I can.'

But instead of heading directly east to Alice Springs with the letters Lasseter gave him to hand to the Government Resident, Johns decided to veer south-east, arriving at Hermannsburg Mission. The mission wired the Government Resident about how Lasseter was now acting alone, but Johns made no attempt to head for Alice Springs. He opted for a leisurely return. 'I stayed at Hermannsburg about a fortnight,' he told Hill. 'I think I should have gone back to him, but I was engaged to the company.'

There was a good reason for Johns's reluctance to return to the Alice. According to the superintendent of the Hermannsburg Mission, Pastor Albrecht, Johns told him that, having left Lasseter, he was heading to Alice Springs when 'he decided to open the letter and find out what his boss had written. What he found out was not flattering. In it . . . he [Lasseter] reported the gun incident and asked that his camel man be locked up for having threatened him with a firearm.' When asked by the pastor what he was going to do with this illegally opened letter, he calmly replied: 'I am going to deliver this letter to the authority to which it is addressed, saying I accidentally spilt some tea over it, and since I could not deliver it in a dirty envelope, I put it in a new one.'

According to Albrecht, when Johns returned from where he had left Lasseter: 'In his hand he carried a new .44 rifle which he handed over saying: "You are a J.P., aren't you?" When we confirmed this, he continued: "I request you to keep this rifle until the police have been informed and given direction as to what is to be done with it. I have taken it from Lasseter.'" Yet Fred Blakeley didn't mention Lasseter having a rifle – .44 calibre or not – when he'd left with Johns. The expedition leader said he had given Lasseter his own revolver. Further, Albrecht wasn't to know that the two rifles carried with the expedition were not the heavy .44 calibre, they were the more common and much lighter .32-20. Whichever way you looked at it, Johns's behaviour and claims didn't add up.

Johns had seen the writing on the wall. He was an illegal immigrant who had jumped ship, and a German to boot, not the most popular of nationalities, wandering through a nation that had lost 60,000 Australian lives in a world war only 12 years earlier. If he walked back into Alice Springs without Lasseter, and with no alibi, he was a ready-made villain for a suspicious public. Further, he knew Blakeley could testify about him entering the Ilpbilla campsite brandishing a pistol and telling everyone to put up their hands. At the very least he faced deportation.

By presenting Lasseter's letters to the mission's superintendent, Johns was more or less putting himself in the clear. The note gave credence to the idea that the prospector was alive and out operating on his own. Yet, even after eventually returning to Alice Springs, Johns appeared in no rush to head back out to find the lone prospector. In writing of Johns's reluctance to return to Lasseter's aid, Lutheran pastor Philip Scherer's opinion was that 'Johns remained in Alice Springs for the time being and did not return to Lasseter (as he had promised), because he had so meanly betrayed him to the authorities'.

But there may have been more to the Lasseter–Johns story.

In the 1960s, Pastor Albrecht wrote an intriguing account of Johns's movements prior to making his surprise appearance with the CAGE expedition at Ilpbilla. Albrecht described how Johns turned up at the mission in the winter of 1930. They were not on the best of terms as Johns had worked at Hermannsburg previously and, for reasons the pastor does not disclose, had been barred: 'we had to terminate his contract after less than twelve months'. Nevertheless, Johns was urgently wanting to procure camels, for hire or purchase. 'Why did he need camels?' asked Albrecht. 'He told us about a Gold Exploration Company, newly formed in Sydney . . . sending an expedition to Central Australia in a big truck.' Somehow, Johns knew the expedition would be heading to Ilpbilla in the Ehrenberg Ranges, then across the desert to the Petermanns. He argued that the truck would prove useless and from Ilpbilla they would need camels. 'I am going out there to meet the expedition, so that when they are stuck, they will hire me.' 'And this is exactly what happened,' Albrecht recalled. How would Johns have known where the expedition was heading when the expeditioners themselves had no idea?

•

In *The Man from Arltunga*, historian Richard Kimber's biography of Central Australian cameleer and prospector Walter Smith (1898–1990), a curious story is told about the old-time bushman's brush with Lasseter in 1930 while the CAGE expedition was preparing to leave Alice Springs. According to Smith, he met with Lasseter during one of the prospector's mysterious lone walks around the Alice. After swearing Smith to secrecy, Lasseter wanted to enter into a clandestine discussion about procuring riding camels. Smith advised Lasseter he wanted his mate Frank Sprigg involved in this discussion, and a second secret meeting took place, where Lasseter said he had already made preliminary plans with a young German dingo scalper named Paul Johns who had his own camels. Johns's camels weren't quite suitable for long-distance riding and Lasseter proposed 'that part of the fabulous reef was pegged in the names of Walter Smith and Frank Sprigg if they provided the necessary assistance'; and if they could get hold of some cash for Lasseter too. Smith apparently provided two camels, 'Tommy' and 'Darky', for Johns to take with him for Lasseter, while Sprigg obtained some supplies – 50-pound bags of flour, ropes, a billy and a frying pan.

Lasseter outlined what he intended to do. Simply, he was going to double-cross the expedition. The plan was for Johns to keep out of sight until the CAGE party was far west of Alice Springs, somewhere near the border of Western Australia, and then the dogger would simply appear. Lasseter would then somehow orchestrate to leave the expedition and carry on with Johns, using the camels to continue the search to find the lost reef. Smith and Sprigg thought Lasseter was 'playing a strange game', but the lure of the reef was more than enough for them to take part. The plan fell apart when Johns and Lasseter came to blows – Johns heading off for the Hermannsburg Mission and Lasseter remaining with Smith's two camels.

●

Perhaps if Fred Blakeley had known something of Johns's background he might have thought twice about simply handing Lasseter over to him to search for the reef. In the light of what happened, it's worth looking at the Paul Johns' story.

In his lifetime, Albert Paul Johns went by many names, including Albert Paul John, Paul Jones, Paul Aldon and Paul Bruno Petrie. In a draft version of his book *Dream Millions*, Blakeley cites Johns as announcing his name as Jahns. Whatever his name, he seems to have been a man of extraordinary intelligence combined with an even greater degree of 'rat cunning'.

Born in Diesdorf, Germany, on 17 January 1906, Johns claimed that his father was an important military official who was brutal to him. As a child he was regularly beaten with a cane. In 1926, with the desire to leave Germany behind, 20-year-old Johns sailed for England aboard the *Viola*, where he was discovered as a stowaway and refused permission to land at London's East and West India docks. Instead he travelled to South Africa aboard the *Halle* and then to Port Adelaide, where he jumped ship.

The police discovered him hiding at the Stonyfell vineyard near Magill in South Australia. He was arrested but was then allowed to stay, working in a local winery under the care of a German manager. On arrival in Australia, Johns had very little English, but he quickly learned the language while employed doing itinerant work on farms around Mintaro and Clare.

In the late 1920s, Johns befriended another German of a similar age, Carl von Czarnecki, who had also jumped ship in Adelaide, finding his way into the network of South Australia's strong and insular German community. Von Czarnecki had many adventures with Johns in South and Central Australia and years later could recall many of his friend's extraordinary antics. Von Czarnecki was seemingly quite in awe of Johns, and fascinated by his unpredictable conduct.

While ostensibly charming and engaging, Johns's behaviour was often mercurial. If he felt slighted at even the most insignificant thing, Johns was quite capable of exacting revenge. Von Czarnecki recounted a story of a hermit in South Australia who would stare at the two young Germans when they walked by his caravan. Johns was quite affronted by this, whispering to von Czarnecki, 'You wait, I'll teach him not to stare at us.' Days later, Johns and von Czarnecki were hauled before the local police.

Someone had ransacked the caravan, wrenching out all the cupboard drawers and strewing the hermit's belongings outside on the ground. 'Could you imagine anybody doing a thing like that?' Johns asked the sergeant. As von Czarnecki recalled, the policeman 'knew very well who would have most likely have done it. By that time he knew Paul's ways pretty well.'

Von Czarnecki's observations about Johns's idiosyncratic behaviour were concise: 'Paul could do anything like that out of spite. He was very resourceful and could tell fantastic lies.' Retribution wasn't his only skill. 'He was also incidentally an expert pickpocket,' recalled von Czarnecki, 'and could pick any lock.'

Johns also had an unhealthy predilection for brandishing firearms. One evening while eating a roast duck meal with von Czarnecki, Johns became agitated by the presence of a blowfly. He left the room to return with a loaded shotgun; he discharged the weapon at the fly, blowing a hole in the wall and showering the meal in lead pellets and shards of glass.

Petty crime was another specialty. Once when von Czarnecki was with Johns in a store in Clare, while charming the proprietor, Johns leaned forward with his folded overcoat and 'swiped the whole batch of books from off the counter. He had a terrific nerve. And what he didn't know he made up . . . Paul was like that. He could "con" or 'fool anybody,' remembered von Czarnecki.

Yet it seems that the one thing this con man could not abide was being conned by another. He had smelled a like-minded impostor with the visiting Hauptmann von Berlichingen, and thought he smelled another with Lasseter. Both had had their mail torn open by Johns and both were confronted and ridiculed by the master con man himself – von Berlichingen driven off Hermannsburg Mission and Lasseter left to wander the wilds west of Alice Springs.

Johns's time in Australia after Lasseter's demise was initially spent performing itinerant work around Alice Springs and on various mining sites. He applied for, and received, Australian naturalisation in 1933 after renouncing his German nationality. For a while, Ken Vincent Harris and his wife Claire, who operated the Tanami Gold Mine, employed Johns as a general labourer. Harris, a welterweight boxer, who did much in cleaning up the lawless Tanami region to lessen its gun-slinging, 'wild west' reputation, remembered Johns as a 'bragging German'. At one time, he said, Johns boasted that his fight in the desert with Lasseter ended with a particularly violent twist. Johns told Harris that Lasseter 'was mad, and there was no gold', and that the pair 'struggled with a firearm, it discharged and Lasseter was shot'. Johns's story often seemed to vary with differing degrees of excitement each time he retold it, and the addition of a shooting might sound extreme.

However, Lasseter's daughter Ruby wrote that she understood Johns returned from the desert 'to say they had an argument during which he had shot and disabled my father's left arm and left him with some food and water for two camels ...' Away from the newspapers he gloated about the incident. Another of Johns's claims was that the reason he left Lasseter to return to Alice Springs was 'to get him a straight-jacket'.

Johns's reckless behaviour eventually saw him in trouble with the law. He moved interstate regularly, most probably to keep one step ahead of the police. His misdemeanours varied in scale, audacity and location. They included passing valueless cheques in Alice Springs; theft of car registration in Victoria; theft of clothing from a menswear store in Adelaide; illicit dealing in gold in Tennant Creek; car theft in Sydney; and even organising an elaborate interstate car-rebirthing racket in Brisbane (he was found with acids and stamping dyes to create new serial numbers). Together with another like-minded individual, Jack Hicks, he set up a fraudulent mining company in Melbourne, the Toledo Gold Mining Company, which netted a large amount of money before the pair absconded to Sydney, where they were arrested and charged with defrauding public money, false advertising and car theft. Johns was advised by the superintendent of police in Melbourne to leave the state and never return. He spent varying lengths of time in gaol at Alice Springs and in Fanny Bay gaol in Darwin, and did six months' hard labour in Brisbane before being deported in 1937.

•

In 1944, Carl von Czarnecki was picked up and arrested by the US 12th Army Group in Germany, where his name turned up on a list of German undesirables. 'It is understood this man's name was included in a list supplied by Australia to Dominions office of British subjects of German extraction and Nazi sympathies who left Australia for Germany before the war and whose wartime activities were unknown.

'Among his possessions was also found a number of letters in German written to him at Alice Springs from Albert Paul Johns alias Jones, alias Aldon, an un-naturalised German, present whereabouts unknown, who was suspect from a security point of view and believed to be connected with Nazi activities at Alice Springs.' Whatever 'Nazi activities' were taking place in Alice Springs in the mid to late 1930s one can only presume would be on a small scale. Nevertheless, wartime correspondence from the Commonwealth Investigation Branch claims that Johns 'had been a member of the Nazi organization whilst in Australia'.

In the late 1930s, von Czarnecki and his young family moved from Alice Springs to eventually reside in Perth, Western Australia, where Johns became an occasional visitor. Johns and von Czarnecki were taken under the wing of an older couple, Paul and Erna Teske. Erna was the German vice-consul in Western Australia. She also happened to be the state leader of the NSDAP – the National Socialist German Workers' Party, more commonly known as the Nazi Party. Erna Teske certainly took an interest in von Czarnecki's struggling family. Letters held in the National Archive written to her Sydney Nazi counterpart regarding their welfare signed off with a cordial 'Heil Hitler'. Precisely how indoctrinated with Nazi ideals Paul Johns became in the late 1930s is not known, but if it was anything like his friend Carl von Czarnecki, it would be nothing short of alarming.

Not long before the outbreak of World War II, both were back in Germany – Johns deported as an undesirable, and von Czarnecki returning to find a cure for a debilitating skin disease. After war broke out, Special Branch operatives in Australia intercepted mail from von Czarnecki to his wife that was effusive in its praise of the new Germany, one letter reflecting, 'The more I see, the more I am convinced Australia would be much better off if they had a Hitler.' Special Branch reported him as 'open in his regard for Nazi activities'. Security Service interest in von Czarnecki, who was now working for the German News Service as a translator, saw a concerted investigation into his affairs, which then turned up the name of Paul Johns.

The Perth Security Service was keen to trace Johns's and von Czarnecki's movements during the build-up to war. 'It has been established that Czarnecki received a visit from Johns at Narembeen [in Western Australia] in either 1937 or 1938, the exact date is not yet known ... Subsequent inquiry made with a view to establishing the date of Johns's visit to Narembeen has been without result ...'

Von Czarnecki's former employer in Western Australia was brought in for questioning, the interview report outlining: 'Von Czarnecki had spent some time in Alice Springs prior to coming to WA . . . Whilst in Mr Cusack's employ von Czarnecki received a visit from another German from Alice Springs, who had come to Australia with, or about the same time as von Czarnecki. He cannot remember the name of this visitor but states that he heard later that this other German was convicted of a civil offence, dealing with two bank accounts he had . . . Perusal of von Czarnecki's papers, etc. showed a newspaper cutting giving this man's name as Albert Paul Johns.'

But unbeknown to them, Johns had already been picked up by MI5 in London as a saboteur, apparently trained in a 'Nazi instruction school'. Other information attained was that he had two brothers in the Luftwaffe, he'd boasted that he'd participated in the pogrom against the Jews, and that he had most likely been 'a member of the Secret Service'. It was concluded that Johns had arrived in Britain to link up with the British Union of Fascists, with the intent of recruiting those sympathetic to the fascist cause and carrying out operations as a fifth columnist. He was interned for the duration of the war with other members of the British Union of Fascists. A facility on the Isle of Man was as remote a place as any to exile the black-shirted.

At war's end, Johns and his new wife went to Rhodesia and eventually returned to Britain, where he became an antiques dealer, setting up Constance Johns Antiquities in Westerham, Kent.

He remained somewhat elusive to those wanting to question him about his Lasseter days, one historian who'd been looking for him in the 1980s writing, 'So far all efforts to find him have been unsuccessful.' But in his later years he did give the occasional interview, in 1973 're-fashioning' his story to suit his audience . . . as he always did. 'Mr Johns . . . is convinced that Lasseter found gold . . . he has no doubt about gold being in the area where Lasseter perished . . . He speaks of Lasseter with respect and criticizes him only for the fact Lasseter didn't understand the immutable laws of survival in the desert . . . Mr Johns went back to Germany in 1939, didn't like Hitler or his theories and got out hurriedly to England . . .' He seemed to have forgotten telling Phillip Taylor all those years ago that Lasseter was a 'rogue and has no reef to discover'. He also forgot to mention his arrest by MI5 and internment as a saboteur.

36

TAILINGS...

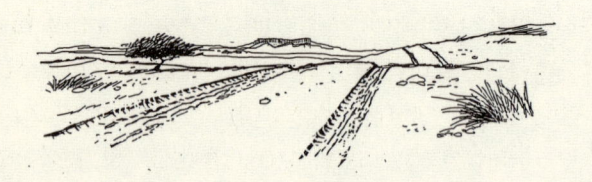

'Success has many fathers but defeat is an orphan', so the saying goes. It was certainly true of the wash-up following the disaster that was the Central Australian Gold Exploration Company. Aside from the enigma of Harold Lasseter, the two questions that arose from the smoking wreckage were: who was responsible for what happened? And what of the gold reef?

There would always be a scapegoat, and perhaps that scapegoat should have been Harold Bell Lasseter. But he was dead, or (depending who you listened to) alive somewhere in the United States.

The CAGE company chairman John Bailey, in his report 'The History of Lasseter's Reef', turned his sights squarely on expedition leader Fred Blakeley, blaming him for the debacle, handing down a damning verdict as though he was heading a royal commission, the union boss finding Blakeley guilty of Lasseter's death. 'The act of the Leader of the Expedition, Mr Blakeley, on the 13th September 1930 to relinquish his position... was a fatal mistake.

He should never have let Lasseter get away from him . . . He is a very straight-forward, honourable person but he displayed a lamentable lack of ability as a leader, and I, as one who has had years of experience of the tropics know that to be a good leader is not easy. Those qualities are in a man's blood . . . he failed as a leader because he did not possess those qualifications. The sad ending of Lasseter must be placed upon Blakeley's shoulders.' A nastier summation would be difficult to imagine.

There is little question that Blakeley was out of his depth as leader of the expedition. He was a turn-of-the-century man dealing with technology 30 years beyond him. He had no experience in mounting a desert campaign of this magnitude and importance. Yet he had unquestionable experience in the far-flung outback and was an accomplished miner. But as it transpired, he was never put there for his talent anyway; as the odious company secretary and John Bailey's son, Ernest, would churlishly remark about Blakeley, 'He was just a nincompoop! . . . It was only because his brother was a Federal minister and we thought this connection might be useful.' Ern Bailey's admission that they were expediently using Fred Blakeley to have access to his brother is distasteful, but hardly surprising.

As the expedition leader, Blakeley never stood a chance. From the outset he found himself surrounded by a core of questionable individuals who didn't fit the bill for participating in any kind of exploration mission. He didn't know, but many had form. John Bailey, the company chairman and AWU boss, had at one time been expelled from the Labor Party for fraudulent behaviour and had been described by his own party as having had 'an unparalleled career of criminal conduct'. Harold Lasseter and Errol Coote were convicted criminals who had served time – Lasseter in a juvenile reformatory for conducting an armed burglary, and Coote in several military prisons for a variety of petty crimes, including theft and wilful escape, resulting in his court martial. Paul Johns, although he hadn't served time yet, would very soon be in the Alice Springs lock-up, then doing hard labour in a Brisbane prison, eventually deported to Germany as an undesirable and then interned in Britain for the duration of the war as a Nazi saboteur.

Blakeley's expedition was stymied from the start, particularly by Lasseter and Coote, who were both running hidden agendas – sometimes separately, sometimes in tandem. Their clandestine, toxic accounts of Blakeley's leadership relayed to the Sydney office quickly white-anted his authority. Blakeley saw the writing on the wall: that he and everyone else had been duped and that he was being set up to carry the blame. He eventually called Lasseter's bluff and followed his whims as to where the gold reef might be in the desert, hoping to give him enough rope, exhausting all his avenues to come up empty-handed. And as far as Blakeley was concerned, Lasseter's inability to turn up any proof of the reef's whereabouts or even its existence vindicated his decision to terminate the search. But the company in Sydney didn't see it like that.

The CAGE board might have labelled Fred Blakeley as a failure, but there was a lot of failure going around far west of Alice Springs in 1930. Lasseter failed to honour his agreement with the company to reveal the location of the reef. Johns failed to honour his agreement with Lasseter to help him. The company failed to support Blakeley as leader. And Errol Coote failed to contribute anything to the expedition other than trouble, expense and catastrophe. In essence, all of these individuals contributed to the collapse of the expedition, and to Lasseter's eventual demise.

Fred Blakeley might have called a halt to operations out in the desert, but he didn't shirk his responsibility in presenting his case to the board as to why. He came to Sydney to blow the whistle. But the company chairman did not read out Blakeley's damning report at the general meeting of shareholders. The last thing John Bailey wanted was a whistleblower.

As it would transpire, Bailey would be no stranger to dubious corporate behaviour when it came to shareholder-funded companies. Less than two years later, John Bailey and Frank Green, a prospector from the Second Expedition, set up the Arnhem Land Gold Development Company, its flotation on the stock market later described as 'constituting one of the biggest scandals in the history of gold mining in Australia'. In Canberra, the independent member for the Northern Territory, Adair Macalister Blain, came out swinging at Bailey. He was fed up with the escalating influx of fraudulent mining ventures pouring into the Territory, instigated by crooked 'promoters' from Sydney and Melbourne. The Arnhem Land Gold Development Company had tricked investors into buying shares and leases in a worthless mine that had produced bogus gold samples. Blain described the company prospectus as 'untrue from end to end', and said 'that at present no-one was making anything out of mining in the Northern Territory except go-getters and scoundrels and the iniquitous Sydney Stock Exchange'.

Blain claimed that Bailey and Green, 'these Labor men', had taken £80,000 of 'promoters' shares out of the company, from a mine that had only ever produced four pounds worth of gold. He concluded that he 'could not find a single statement on a material matter made by the company or its manager which was not true'. In the official history of the Australian Workers Union, Bailey was described as having been 'ultimately cast into the wilderness'. Politically, 'Ballot Box Bailey was a liability in the world of NSW Labor. The Executive Council believed that the Central Branch was mismanaged, something perhaps indicated by lavishing funds on wild prospecting schemes.'

•

There are a lot of 'moving parts' in the Harold Lasseter story, and to try to single out a precise moment as the point of no return for Lasseter is fraught with complications.

It is quite apparent that Lasseter did not want to return to Alice Springs when Blakeley decided to call the expedition quits. It might have been stalling for time in the hope of relocating his reef; fear of being exposed as a fraud and sent to prison; or a secret agenda to leave the expedition to link up with the mysterious Johansen. Blakeley admitted that it was his idea for Lasseter to find a way to simply disappear, and that he suggested he team up with the German dogger Paul Johns when they returned to Ilpbilla. It was a way out for Lasseter, certainly. It would be the last time the expedition members would lay eyes on Harold Lasseter. Even though Blakeley had reservations about leaving him with the young dogger, his concerns were in regard to the German's survival ability rather than his then-unknown erratic temperament.

In hindsight, perhaps Johns might not have been the wisest choice for someone to venture out alone with into the desert. If what Walter Smith said was true and Johns had been hired by Lasseter to suddenly materialise with a brace of camels, perhaps to allow him the opportunity to hook up with Johansen at Lake Christopher, then it is possible that this fragile deal between two fractious partners broke down. But one thing is certain: that Johns and Lasseter were involved in a violent altercation, from which only Johns returned. If the diary recovered by Bob Buck and eyewitness accounts from local Aborigines are genuine, then Lasseter died a lonely death out at Irving Creek at the end of January 1931. But as the diary records, Lasseter said that Johns gave him his 'word of honour' he would return – yet instead of making for Alice Springs as a matter of urgency, Johns proceeded to Hermannsburg Mission, where he remained for two weeks. He had illegally opened Lasseter's letter addressed to the Government Resident, which described the fight incident and demanded Johns be placed under arrest. As Fred Blakeley understood it, the charge against Johns was for 'some sort of violent assault'.

Even though the authorities had been notified about Lasseter's predicament from the mission's wireless, Johns took his time to wend his way to the Alice, then handed in Lasseter's resealed letters and did nothing for a further two weeks. As Pastor Philipp Scherer put it, 'Johns remained in Alice Springs for the time being and did not return to Lasseter (as he had promised), because he [Lasseter] had so meanly betrayed him to the authorities.' Spite is a recurring theme in the Paul Johns story. As his friend Carl von Czarnecki once observed, 'Paul could do anything like that out of spite. He was very resourceful and could tell fantastic lies.' It was Errol Coote who had remarked of Lasseter, 'If he has brought us out on a wild goose chase we'll give him his rations and water and make him walk back.' Perhaps Johns, who possessed a reputation for exacting vengeance on those who crossed him, left Lasseter – of whom Johns said he 'did not impress me as a man who knew the country, but rather as one who had read about it' – in the desert to perish.

It is difficult to appreciate solely from the written word of almost a century ago the nuances of what it was genuinely like in the field in 1930. There is a lot recorded, much first hand, about what transpired, but it is impossible to capture the tenor and undercurrents of gut feelings on a rolling expedition like this. For every theory there is as to what happened to Harold Lasseter, there is another. And then another.

To short-circuit the threat of culpability, the key players scrambled to write their accounts of what happened during the ill-fated CAGE expedition. Paul Johns gave his statement to historian Ernestine Hill in 1932, an austere account of his movements while involved with the expedition. As Johns was known for boasting about his actions, it is unlikely that the document gives the full story. In 1934, Errol Coote, a journalist, published *Hell's Airport*, which detailed his experiences as the expedition's official pilot, and later as the replacement leader. While entertaining, the book is laced with thinly veiled criticisms of almost everyone involved, before culminating with his stupefyingly irresponsible flight to Ayers Rock, proving and accomplishing nothing. In 1937, Fred Blakeley wrote the manuscript *Dream Millions*, which would not be published until 1972, ten years after his death. In the book Blakeley attempts to justify his decisions up until the point he was sacked, and then embarks on a robust campaign, venting his anger. John Bailey wrote his report 'The History of Lasseter's Reef', laying the blame squarely on Fred Blakeley. His summary was that 'Lasseter was a liar and a fraud'. The manuscript is held in the Mitchell Library along with some other items donated by Bailey. The bulk of the CAGE documents were burned soon after the expedition finished.

As for the gold reef, at the point of publication no-one has rediscovered it – yet. And not for want of trying. It would be reasonable to assume that at any given time since 1930, there has always been someone poring over a map of Central Australia, working away with whatever was the latest in technology at the time – from a nib pen with a prismatic compass and a Douglas protractor to GPS – attempting to be the first person to crack the mystery. Yet the odds of the reef's existence grow slimmer as each year passes. The more people have tramped around Central Australia, and the better technology has become in trying to locate it, the further away the reef seems from our reach than it was back in the days of the CAGE expeditions.

It is impossible to ascertain just how many forays since 1931 have set off for the wilds in what has become known as 'Lasseter country'. The years immediately following Lasseter's demise saw what seemed like a stampede of expeditions headed by people who *knew* where the reef was, and who on occasion would return brandishing questionable ephemera, supposedly belonging to the prospector himself, that they had discovered under a fire or in a sand dune.

Common sense tells us Lasseter's reef does not exist. It was a lie, or a hoax, or a figment of the old prospector's imagination. Even in 1930, his story had the ring of a well-worn fable about the lust for lost gold and the disaster that would ensue in searching for it. Every few years a story would appear in a local newspaper about two mates who knew where the reef was; they'd picked up some lost snippet of information or deciphered secret coordinates; their photograph depicting them standing alongside a four-wheel drive, the bonnet strewn with maps and documents, ready to turn the key and head off to locate the lost El Dorado. But time would pass and the follow-up story about how they found the reef never seemed to materialise.

The search for Lasseter's reef was always going to be a gamble. Herbert Gepp's report to the government even said so. Gamblers know the phenomenal odds stacked against them in horse racing or playing poker machines, or even buying a lotto ticket, as being millions to one, and common sense should tell us it is a waste of time and money to even try. But despite the preposterous odds, people do – on occasion – win. And this spurs us on.

What if Lasseter's gold reef was just over that next rise? What if . . .

It is the irrational desire to believe in the unbelievable.

Perhaps the key to locating Lasseter's fabulous gold reef is not to look for it at all. It is clearly not meant to be found by hurling manpower and money and technology at the mystery in an attempt to force the land to surrender its location. If the reef is ever to be discovered, it will reveal itself in its own time and in its own way – when someone stumbles upon it by sheer accident.

After all, that's what Lasseter did.

AFTERWORD

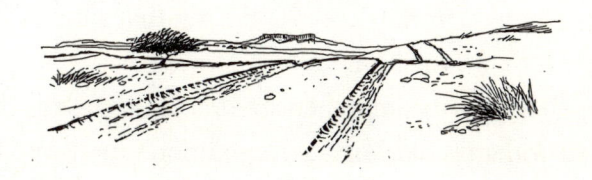

The dust settled quickly for those involved in the Lasseter story. That there was no official inquiry or investigation as to what happened in the hunt for Lasseter's gold beggars belief. It's a wonder there wasn't a Royal Commission – very little seemed above board. Australian Workers Union boss John Bailey had squandered tens of thousands of pounds of union members' funds on a publicly subscribed expedition because he'd been taken in by an unlikely story. A federal minister – Arthur Blakeley – came under relentless attack from the opposition for seemingly pulling strings to mobilise RAAF aircraft to bail out the expedition's line-up of inept flyers. No other expeditions were given such treatment. And further, it was constantly brought to the minister's attention his brother was the expedition leader.

Perhaps there was a reason why there was not an investigation – wherever a stone was unturned, there would always be some unexpected, unsavoury entanglement with opportunistic politicians, businessmen and union officials. Even the seemingly altruistic and magnanimous A. G. Hebblewhite who kindly 'loaned' the Thornycroft truck was in reality something of a trade union apparatchik – becoming the vociferous Administrator of the People's Union.

But more importantly, an investigation should have taken place into what happened to Harold Lasseter, who had been abandoned in the wilderness for ninety days before he ultimately perished. Despite the proliferation of aircraft and trucks and camel trains traipsing around Central Australia during that time, no-one seriously went looking for him.

Events that have incomprehensible endings, such as that of the Central Australian Gold Mining Expedition's foray into Centralia during 1930, will undoubtedly attract conspiracy theorists. Fred Blakeley was convinced Lasseter had made a masterful escape from Central Australia and was very much alive – a theory supported by many and backed by his correspondent Mrs Nellie Edwards writing to him in the 1950s, 'I sware (sic) beyond any doubt at all the man I met . . . was Harold Bell Lasseter.'

Yet generally, there is a simple answer for the inexplicable – that there is no conspiracy at all. What transpired had been the result of some human error or simply bad luck.

But not quite so in the case of the search for Lasseter's gold reef. There is no question various expedition and company members were operating hidden agendas and making secret plans, concealing information from other colleagues and shareholders and simultaneously feeding them false material. Individual, personal conspiracies that eventually – and inevitably – collided, leaving the entire operation as a smoking wreck with Harold Lasseter very likely dead as a result.

Lasseter's story is in effect a morality tale. It has all the power and wrathful message of some fire-and-brimstone Old Testament parable. If you were to try to pinpoint precisely what Lasseter's story is about, it would come down to one word: greed. In the aftermath of what transpired, this was often euphemistically termed as something like 'the lure of gold' – but straight-out greed was simply the grubby, expedient driving force for a cabal of opportunists to become rich whatever it took. In fairness, had there not been the unexpected calamity of the Great Depression it is very likely no-one would have genuinely gone to the lengths John Bailey did in setting up an expedition such as the CAGE enterprise. But unfortunately for everyone – particularly for Lasseter himself – the prospector materialised in what we would term today as 'a perfect storm': the Great Depression, a flailing government, the desperate need for quick money. For decades Lasseter had been throwing incongruent ideas one after the other at the federal government – and for the one and only time – they actually took an interest in what he had to say, ultimately sealing his fate.

It is amazing that after so many years, missing pieces of the Lasseter puzzle still appear. In the time of writing this book, Luke Walker's remarkable documentary *Lasseter's Bones* solved the mystery as to who the elusive Johanson was.

And now some other curious, missing elements have emerged: one of the hand-written letters supposedly found on Lasseter's body by Bob Buck, Buck's telegram to John Bailey dated 27 April 1931 announcing his grisly discovery, and an envelope signed by Lasseter and opened by his solicitor on 18 May 1931. The Government Resident in Alice Springs, Vic Carrington, posted the letters to The Department of Home Affairs, where they were carefully transcribed – the typewritten account is held in the National Archive of Australia. The original letters – written in lead pencil – mysteriously disappeared. It was thought perhaps they were purchased from Lasseter's widow by the author Ion Idriess.

In May 2015 I was contacted by journalist Fiona Smith of the *Australian Financial Review*, who told me she had recently held one of the letters retrieved from Lasseter's body. This letter, not seen since the 1930s, was in the possession of a family friend with whom I was able to make contact. It is not known precisely how the letter and telegram ended up as part of the estate left by Richard and Jill Stalley of Sydney – there is some conjecture among family members that it may have been passed down from Richard's father. The letter, frail and tattered, matches the typewritten transcription in the National Archive word for word, to the point where words no longer legible along the letter's deteriorated edges correspond with the broken wording of the transcript. And of Bob Buck's wonderfully preserved telegram announcing the discovery of Lasseter's remains – along with sundry items found at the scene such as the revolver – this is also a truly significant find. Perhaps, in time, Lasseter's past will give up other secrets – a reef of gold stretching as far as the eye can see . . .

ACKNOWLEDGEMENTS

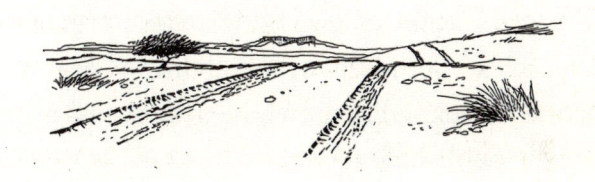

In this book I have tried to utilise as many first-person accounts as I could find. Much of the Lasseter story that has been handed down through the years is a bit like grandfather's axe – it's the same axe but has had four new handles and five new heads.

The most renowned work written about Harold Lasseter and his gold reef is Ion Idriess's *Lasseter's Last Ride,* first published in 1931. Yet while it remains an Australian classic, I have been circumspect in using *Lasseter's Last Ride* as a key source. Idriess never met Lasseter and the book is, ultimately, fiction intertwined with fact.

The two key manuscripts I have used to drive the narrative were accounts authored by expedition leader Fred Blakeley in his book *Dream Millions* written in 1937 and the expedition pilot Errol Coote's book *Hell's Airport* written in 1934. Both were excellent writers – yet as I understood it, the two books were apparently incompatible. The stories they tell were markedly different, presumably because Blakeley and Coote were perpetually at loggerheads. However, I found the reputed conflict between the two books to be unsubstantiated – *Dream Millions* and *Hell's Airport*, in my opinion, complement each other. Certainly there are differences regarding many aspects of their stories, but by and large they correlate with what took place and the general feeling amongst the expedition party regarding Harold Lasseter. Fred Blakeley's unpublished manuscript held in the Mitchell Library is particularly illuminating – the edited, published version seems to have been somewhat emasculated. A lot of Blakeley's vitriol is left in the archive – perhaps because several of the key expedition members were still alive when it was eventually published.

There are many other publications I have used to guide me through the story and would like to recommend – I am particularly grateful for the work published by both professional and amateur historians. Yet the Lasseter story in its entirety has been told rarely. Billy Marshall-Stoneking's book *Lasseter in Quest of Gold*, first published in 1985, is a particularly comprehensive account of what took place. Marshall-Stoneking's time living with Aboriginals in the Northern Territory gives his story added dimensions. Murray Hubbard's *The Search for Harold Lasseter* is an extraordinary book – a sort of rolling adventure as he travels around Australia piecing together Lasseter's life, and Max Cartwright's *Ayres Rock to the Petermanns* is another fascinating read. Richard Kimber's *The Man from Arltunga*, Walter Gill's *Petermann Journey* and Alan Powell's *Far Country* give marvellous insights into what life in the frontier world of Central Australia was like.

A special mention goes to the website Lasseteria.com – an extraordinary resource assembled and maintained by Robert Ross, someone truly dedicated to keeping the Lasseter legend alive. It would be impossible to search the internet for any information on Harold Lasseter without some link to Lasseteria appearing. The colossal wealth of detail found here is staggering – to the point of being intimidating! For anyone who has been bitten by the Lasseter bug, Lasseteria is an entire world dedicated to the Harold Lasseter phenomenon. I would also like to thank Robert for sending me the 1954 interview with Fred Blakeley from *People* magazine.

The State Library of NSW, as always, was exceptionally helpful – many thanks to the 'cavalier' Maggie Patton, Kevin Leaman, Cheya Cootes and especially Vanessa Bond who once again pulled a rabbit out of a hat in regard to finding for me one elusive quote buried away within their archives.

To ocean-going seafarers Campbell Reid, Piers Akerman and Roger Coombs who were able to relate to me the sensation of loneliness and despair when away at sea, how to use a sextant, and Piers's marvellous advice on *Brown's Nautical Almanac*.

To firearms expert Robert Coombs for his knowledge and advice on weapons and calibres popularly used in Australia during Lasseter's time, and the wonderful Laurie Wallace of the Morse Codian Society who gave me the most extraordinary lesson about the history of Morse in Central Australia. Thanks also to the ladies at the Old Telegraph Station in Alice Springs.

To my good friend Richard Farrant – a member of the Military Vehicle Trust in the United Kingdom – who came to Australia to advise me on the Thornycroft A3. The truck – to my mind – was every bit as important a character within the story as any of the expedition members. Owning two 1920s British trucks and having driven a World War II six-wheel-drive truck from Alice Springs to Darwin, I was particularly keen for the reader to appreciate just what a terrible undertaking the drive to find the reef was.

Many thanks go to great mates Geoff Simmons and Lynne Brown – two specialist mechanics who have an unassailable knowledge of pre-war vehicles. To a dear friend, Ross Cattarall, whose memories of growing up as a boy in the outback have been of invaluable assistance.

To Ken Harris, whose crystal clear memories of his father's experiences with Paul Johns at the Tanami Gold Mine really helped flesh out my picture of him. And a special thanks to John O'Brien, Ken's great mate – a man with a mind like a steel trap.

To Trevor Bartlett – who is so generous with his time and enthusiasm in hunting through the jungle of family trees, and to Tim Creer who presented me with his copy of David Hill's *Gold!* – my introduction to Harold Lasseter.

A special thank you to Graham Golding who runs the Universal Books store in Taralga NSW – someone who has the phenomenal knack of procuring hard-to-find Australian literature. At any given time Graham could produce some rare and obscure volume about Central Australia published in the 1930s that really helped in setting the scene. A wonderful bookshop.

Thanks to Major General Mark Kelly whose experience and wisdom is so highly valued – particularly on the difficulties in handling recalcitrant groups such as the CAGE expedition.

To the Stalley family for letting me see their incredible treasure: Lasseter's last letter and Bob Buck's telegram. Thank you for letting me reproduce them in this book for Australia to see. And a grand thank you to my cousin Fiona Smith of the *Australian Financial Review* who organised this extraordinary part of the Lasseter story to be revealed.

At the *Daily Telegraph*, many thanks for the patience of my editor Paul (Boris) Whittaker, deputy editor Ben English and to Brad Clifton and Saturday editor Jeni O'dowd.

To the editor of the *Sunday Telegraph* Mick Carroll, deputy editor Claire Harvey (who described to me in detail her interview with Errol Coote's widow, Alison) and to Lillian Saleh.

To former journalist David Jensen who interviewed Paul Johns in the early 1970s. David, then a 'stringer' in the UK, picked up the story of Johns as a general release 'on the wire' and interviewed him over the telephone – completely unaware of his intriguing past.

To Stuart Duncan and his wife Debbie, who now reside on a bleak island somewhere off the west coast of Scotland – about as far from Central Australia as imaginable, thanks, mate, for your never-flagging support.

To a great friend, Associate Professor of History Andrew Moore, whose knowledge of politics through this period was invaluable. And to the wonderful Meredith Burgmann – about the clearest thinker I know – who can always come up with fresh perspectives. To the Australasian Primates – every one of you.

Thanks also to John and Bronwyn Sullivan, David and Katherine Medina, David and Catherine Mitchell, Matt and Jeanette Simpson, Liza Duncan, Michelle Marshall, Martha Graham, Gabby Keith, Glenn Hall, Mick and Nick, Laurie and Jimmie Dean.

Many thanks to my agent, James Laurie – a true gentleman, with whom I'm looking forward to catching up.

And a special thank you to my publisher Matthew Kelly, who has persevered with me throughout the production of this book. There were times I was flummoxed as to how to go about writing it – it was a surprisingly complex story that didn't quite have a beginning and certainly did not have an end. A lot of the information about what happened appeared to be hearsay – most of it conflicting – and the number of tangents and sub-plots and theories and dead ends were almost overwhelming. I then realised as to why the story in its entirety had been so rarely told. Matthew stuck with it – in an 'I'll get this book out there if it kills me' kind of way – a dogmatic approach that pushed me forward, seeing it through to the end. Thank you, Matthew – I am eternally grateful for your patience.

And to beautiful Tanya and Oliver Brown, aged 8, who make every single day such a pleasure.

SOURCES

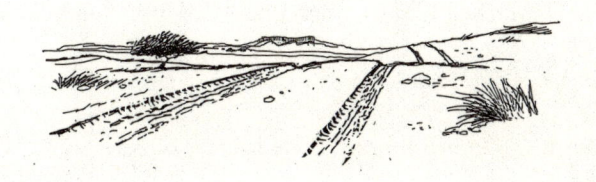

Since Harold Lasseter's demise in 1931 there has been no shortage of books written about the prospector and what transpired in the search for the gold reef. Below are some of the books and tangential publications I've used.

Note that in writing this book I refrained from seeking a meeting with Harold Lasseter's son, Robert. Early on, my friend Piers Akerman asked me whether I would speak to Bob Lasseter – something to which I had given serious thought and ultimately decided against. At the time of writing Bob Lasseter would be ninety-one years of age and has had a lifetime of journalists and writers, documentary-makers, treasure hunters and those obsessed with Lasseter hassling him for his opinions, advice and knowledge. Bob Lasseter was six when his father failed to return from the red centre – and since then has conducted many expeditions into 'Lasseter Country' looking for his father's lost reef. In the remarkable 2012 documentary *Lasseter's Bones*, film-maker Luke Walker's focus shifts onto Robert Lasseter and his quest to exonerate his father's name. What you see is a warm, intelligent and dedicated person whose beliefs are in no doubt. I saw in Robert something of my own father – a gentleman – and decided to leave Robert and his wife in peace.

A Son of the 'Red Centre' – Kurt G. Johannsen, edited by Daphne Palmer, Gecko Books, Adelaide, 1992.
Ayers Rock to the Petermanns: Legend of Lasseter – Max Cartwright, Alice Springs, 1991
Bailey Papers – Mitchell Library, State Library of New South Wales, Sydney.
Black Kettle and Full Moon – Geoffrey Blainey, Penguin Books, Melbourne, 2003.
Blakeley Papers – Mitchell Library, State Library of New South Wales, Sydney.
Brown Men and Red Sand – Charles P. Mountford, Robertson and Mullens, Melbourne, 1948.
C.A.G.E. Papers – National Archive of Australia, Canberra.
Far Country: A Short History of the Northern Territory – Alan Powell, Melbourne University Press, Melbourne, 1982 (1996 edition).
Gold! The Fever That Forever Changed Australia – David Hill, William Heinemann, Sydney, 2010.
Hard Liberty: A Record of Experience – Frederick Blakeley, George G. Harrap, London, 1938.
Hell's Airport and Lasseter's Lost Legacy – Errol Coote, Investigator Press, Adelaide, 1934 (1981 edition).
Here's Luck – L. W. Lower, Angus & Robertson, Sydney, 1930.
Historic Military Vehicles Directory – Bart H. Vanderveen, Battle of Britain Prints International, London, 1989.
Lasseter Demystified & Two German Rouseabouts – P. A. Scherer, Tanunda, 1996.
Lasseter's Bones – A Documentary by Luke Walker, Scribble Films, 2012.
Lasseter's Dream of Millions: New Light on the Lost Gold Reef – Fred Blakeley, edited by Frances Wheelhouse and Mary Mansfield, Transpereon Press, Sydney, 1972 (1984 edition).
Lasseter's Dream Millions – unpublished manuscript, Mitchell Library, State Library of New South Wales.
Lasseter In Quest of Gold – Billy Marshall-Stoneking, Hodder & Stoughton, Sydney, 1985 (1994 edition).
Lasseter's Last Ride – Ion L. Idriess, Angus & Robertson, Sydney, 1931 (1950 edition).

'Lasseter's Last Ride' in *Ion Idriess's Greatest Stories* – Ion L. Idriess, Angus & Robertson, Sydney, 1931 (1986 edition).
Lasseteria.com – An Encyclopedia on Harold Lasseter and Lasseter's Reef, Robert Ross.
Lost Explorers – Ed Wright, Pier 9, Sydney, 2008.
Nazi Dreamtime: Australian Enthusiasts for Hitler's Germany – David S. Bird, Australian Scholarly, Melbourne, 2012.
Olof's Suitcase: Lasseter's Reef Mystery Solved – Chris Clark, Echo Books, Geelong, 2015.
Outback in Australia – Walter Kilroy Harris, Garden City Press, Letchworth, 1913.
Paul Johns Statement – Ernestine Hill, State Library of New South Wales, Sydney, 1932.
Petermann Journey – Water Gill, Angus & Robertson, Sydney, 1968.
Spinifex Walkabout: Hitchhiking in Remote North Australia – Coralie and Leslie Rees, George. G. Harrap, Sydney, 1953.
The Australian Geographic Book of the Red Centre – Jenny Stanton, Barry Skipsey, Australian Geographic, Sydney, 2005.
The Explorers – edited by Tim Flannery, Text Publishing, Melbourne, 1998.
The Man from Arltunga – Richard Kimber, Hesperian Press, Perth, 1996.
The Man from Oodnadatta – R. B. Plowman, Angus & Robertson, Sydney, 1933.
The Search for Harold Lasseter: The True Story of the Man behind the Myths – Murray Hubbard, Angus & Robertson, Sydney, 1993.
Uluru: An Aboriginal History of Ayers Rock – Robert Layton, Aboriginal Studies Press, Canberra, 1986.

hachette
AUSTRALIA

If you would like to find out more about Hachette Australia, our authors, upcoming events and new releases you can visit our website, Facebook or follow us on Twitter:

hachette.com.au
twitter.com/HachetteAus
facebook.com/HachetteAustralia